In Vitro Cultivation
of Animals Cells

BOOKS IN THE BIOTOL SERIES

The Molecular Fabric of Cells
Infrastructure and Activities of Cells

Techniques used in Bioproduct Analysis
Analysis of Amino Acids, Proteins and Nucleic Acids
Analysis of Carbohydrates and Lipids

Principles of Cell Energetics
Energy Sources for Cells
Biosynthesis and the Integration of Cell Metabolism

Genome Management in Prokaryotes
Genome Management in Eukaryotes

Crop Physiology
Crop Productivity

Functional Physiology
Cellular Interactions and Immunobiology
Defence Mechanisms

Bioprocess Technology: Modelling and Transport Phenomena
Operational Modes of Bioreactors

In vitro Cultivation of Micro-organisms
In vitro Cultivation of Plant Cells
In vitro Cultivation of Animal Cells

Bioreactor Design and Product Yield
Product Recovery in Bioprocess Technology

Techniques for Engineering Genes
Strategies for Engineering Organisms

Principles of Enzymology for Technological Applications
Technological Applications of Biocatalysts
Technological Applications of Immunochemicals

Biotechnological Innovations in Health Care

Biotechnological Innovations in Crop Improvement
Biotechnological Innovations in Animal Productivity

Biotechnological Innovations in Energy and Environmental Management

Biotechnological Innovations in Chemical Synthesis

Biotechnological Innovations in Food Processing

Biotechnology Source Book: Safety, Good Practice and Regulatory Affairs

BIOTECHNOLOGY BY OPEN LEARNING

In Vitro Cultivation of Animal Cells

PUBLISHED ON BEHALF OF :

Open universiteit and **University of Greenwich (formerly Thames Polytechnic)**

Valkenburgerweg 167
6401 DL Heerlen
Nederland

Avery Hill Road
Eltham, London SE9 2HB
United Kingdom

Butterworth-Heinemann

iv

Butterworth-Heinemann Ltd
Linacre House, Jordan Hill, Oxford OX2 8DP

A member of the Reed Elsevier group

OXFORD LONDON BOSTON
MUNICH NEW DELHI SINGAPORE SYDNEY
TOKYO TORONTO WELLINGTON

First published 1993

© Butterworth-Heinemann Ltd 1993

British Library Cataloguing in Publication Data
A catalogue record for this book is
available from the British Library

Library of Congress Cataloguing in Publication Data
A catalogue record for this book is
available from the Library of Congress

ISBN 0 7506 0555 3

Composition by University of Greenwich
(formerly Thames Polytechnic)
Printed and Bound in Great Britain

The Biotol Project

The BIOTOL team

OPEN UNIVERSITEIT, THE NETHERLANDS
Prof M. C. E. van Dam-Mieras
Prof W. H. de Jeu
Prof J. de Vries

UNIVERSITY OF GREENWICH (FORMERLY THAMES POLYTECHNIC), UK
Prof B. R. Currell
Dr J. W. James
Dr C. K. Leach
Mr R. A. Patmore

This series of books has been developed through a collaboration between the Open universiteit of the Netherlands and University of Greenwich (formerly Thames Polytechnic) to provide a whole library of advanced level flexible learning materials including books, computer and video programmes. The series will be of particular value to those working in the chemical, pharmaceutical, health care, food and drinks, agriculture, and environmental, manufacturing and service industries. These industries will be increasingly faced with training problems as the use of biologically based techniques replaces or enhances chemical ones or indeed allows the development of products previously impossible.

The BIOTOL books may be studied privately, but specifically they provide a cost-effective major resource for in-house company training and are the basis for a wider range of courses (open, distance or traditional) from universities which, with practical and tutorial support, lead to recognised qualifications. There is a developing network of institutions throughout Europe to offer tutorial and practical support and courses based on BIOTOL both for those newly entering the field of biotechnology and for graduates looking for more advanced training. BIOTOL is for any one wishing to know about and use the principles and techniques of modern biotechnology whether they are technicians needing further education, new graduates wishing to extend their knowledge, mature staff faced with changing work or a new career, managers unfamiliar with the new technology or those returning to work after a career break.

Our learning texts, written in an informal and friendly style, embody the best characteristics of both open and distance learning to provide a flexible resource for individuals, training organisations, polytechnics and universities, and professional bodies. The content of each book has been carefully worked out between teachers and industry to lead students through a programme of work so that they may achieve clearly stated learning objectives. There are activities and exercises throughout the books, and self assessment questions that allow students to check their own progress and receive any necessary remedial help.

The books, within the series, are modular allowing students to select their own entry point depending on their knowledge and previous experience. These texts therefore remove the necessity for students to attend institution based lectures at specific times and places, bringing a new freedom to study their chosen subject at the time they need and a pace and place to suit them. This same freedom is highly beneficial to industry since staff can receive training without spending significant periods away from the workplace attending lectures and courses, and without altering work patterns.

SOFTWARE IN THE BIOTOL SERIES

BIOcalm interactive computer programmes provide experience in decision making in many of the techniques used in Biotechnology. They simulate the practical problems and decisions that need to be addressed in planning, setting up and carrying out research or development experiments and production processes. Each programme has an extensive library including basic concepts, experimental techniques, data and units. Also included with each programme are the relevant BIOTOL books which cover the necessary theoretical background.

The programmes and supporting BIOTOL books are listed below.

Isolation and Growth of Micro-organisms
Book: *In vitro* Cultivation of Micro-organisms
        Energy Sources for Cells

Elucidation and Manipulation of Metabolic Pathways
Books: *In vitro* Cultivation of Micro-organisms
        Energy Sources for Cells

Gene Isolation and Characterisation
Books: Techniques for Engineering Genes
        Strategies for Engineering Organisms

Applications of Genetic Manipulation
Books: Techniques for Engineering Genes
        Strategies for Engineering Organisms

Extraction, Purification and Characterisation of an Enzyme
Books: Analysis of Amino Acids, Proteins and Nucleic Acids
        Techniques used in Bioproduct Analysis

Enzyme Engineering
Books: Principles of Enzymology for Technological Applications
        Molecular Fabric of Cells

Bioprocess Technology
Books: Bioreactor Design and Product Yield
        Product Recovery in Bioprocess Technology
        Bioprocess Technology: Modelling and Transport Phenomena
        Operational Modes of Bioreactors

Further information: Greenwich University Press,
University of Greenwich, Avery Hill Road, London SE9 2HB.

Contributors

AUTHORS

Dr D. Gor, Royal Free Hospital and School of Medicine, London, NW3, UK

Dr E. Lucassen, University of Greenwich, Woolwich, UK

EDITORS

Dr C. K. Leach, De Montfort University, Leicester, UK

Professor M. C. E. van Dam-Mieras, Open universiteit, Heerlen, The Netherlands

SCIENTIFIC AND COURSE ADVISORS

Professor M. C. E. van Dam-Mieras, Open universiteit, Heerlen, The Netherlands

Dr C. K. Leach, De Montfort University, Leicester, UK

ACKNOWLEDGEMENTS

Grateful thanks are extended, not only to the authors, editors and course advisors, but to all those who have contributed to the development and production of this book. They include Mrs A. Allwright, Miss K. Brown, Mrs A. J. Liney, Miss J. Skelton, Mrs S. Smith and Professor R. Spier.

The development of this BIOTOL text has been funded by **COMETT, The European Community Action Programme for Education and Training for Technology**. Additional support was received from the Open universiteit of The Netherlands and by University of Greenwich (formerly Thames Polytechnic).

Contents

How to use an open learning text
Preface

1 Introduction to animal tissue culture
E. Lucassen   1

2 Outline of the key techniques of
animal cell culture
D. Gor   21

3 Animal cell culture media
D. Gor   45

4 Characterisation of cell lines
E. Lucassen   73

5 Preservation of animal cell lines
D. Gor   99

6 Oncogenes and cell transformation
E. Lucassen   115

7 Hybridomas
D. Gor   133

8 Large scale animal cell cultures
D. Gor   153

9 Three dimensional culture techniques
E. Lucassen   185

10 Genetic engineering of animal cells
E. Lucassen   205

Responses to SAQs   239
Suggestions for further reading   263

How to use an open learning text

An open learning text presents to you a very carefully thought out programme of study to achieve stated learning objectives, just as a lecturer does. Rather than just listening to a lecture once, and trying to make notes at the same time, you can with a BIOTOL text study it at your own pace, go back over bits you are unsure about and study wherever you choose. Of great importance are the self assessment questions (SAQs) which challenge your understanding and progress and the responses which provide some help if you have had difficulty. These SAQs are carefully thought out to check that you are indeed achieving the set objectives and therefore area very important part of your study. Every so often in the text you will find the symbol Π, our open door to learning, which indicates an activity for you to do. You will probably find that this participation is a great help to learning so it is important not to skip it.

Whilst you can, as an open learner, study where and when you want, do try to find a place where you can work without disturbance. Most students aim to study a certain number of hours each day or each weekend. If you decide to study for several hours at once, take short breaks of five to ten minutes regularly as it helps to maintain a higher level of overall concentration.

Before you begin a detailed reading of the text, familiarise yourself with the general layout of the material. Have a look at the contents of the various chapters and flip through the pages to get a general impression of the way the subject is dealt with. Forget the old taboo of not writing in books. There is room for your comments, notes and answers; use it and make the book your own personal study record for future revision and reference.

At intervals you will find a summary and list of objectives. The summary will emphasise the important points covered by the material that you have read and the objectives will give you a check list of the things you should then be able to achieve. There are notes in the left hand margin, to help orientate you and emphasise new and important messages.

BIOTOL will be used by universities, polytechnics and colleges as well as industrial training organisations and professional bodies. The texts will form a basis for flexible courses of all types leading to certificates, diplomas and degrees often through credit accumulation and transfer arrangements. In future there will be additional resources available including videos and computer based training programmes.

Preface

Understanding the properties and behaviour of animal cells, especially at the molecular and genetic levels, is essential to the fulfilment of aspirations to improve the prevention, diagnosis and treatment of diseases. This understanding is also of fundamental importance in evaluating the likely consequences on animal health of changes in our environment especially arising from pollution and it opens up new opportunities in animal breeding programmes. It is not surprising, therefore, that considerable resources are invested in the study of animal cells. This investment has been rewarded by the considerable achievements of contemporary biotechnology in providing new products and procedures in disease treatment and management. New vaccines, specific animal proteins such as interferons, blood factors and hormones, monoclonal antibodies for use as diagnostics and therapeutics and gene probes as diagnostic tools are but a few of the many new and important products.

Underpinning our understanding of animal cell biology and, thereby, providing the basis for the production and design of new processes and products, has been a developing capability to cultivate animal cells *in vitro*. On a laboratory scale such cultures provide materials and systems that are accessible to study whilst cultivation of animal cells on an industrial scale enables the production of sufficient quantities of animal cell-derived materials to be of real value to society. Those engaged in this area believe that merely a start has been made and there is much more that could be achieved. This text is designed to describe the principles and techniques of animal cell cultivation so that others may contribute to this important and expanding area of endeavour.

The text begins with a general description of the principles of animal cell cultivation and introduces the key terms used in this area of activity. This general introduction is followed by a chapter dealing with the central techniques used in cell culture including the selection of sources, explant techniques, tissue disaggregation, techniques for monitoring cultures and the routine maintenance of cell cultures. The important issues of media design and preparation are dealt with in Chapter 3. The themes of monitoring cultures and quality assurance are developed further in Chapter 4 which explains the procedures used to characterise cell lines. The long term preservation of cells and the importance of cell banks are described in Chapter 5.

Animal cells may be broadly divided into two categories; transformed and untransformed. These two groups are quite distinct in terms of growth requirements and growth characteristics. The growth characteristics of transformed cells make these cells attractive for large scale cultivation. The nature of the transformation process and the importance of proto-oncogenes and oncogenes involved in the regulation of cell division, are described in Chapter 6.

In Chapter 7, we examine hybridoma technology in which the growth characteristics of one cell line and the specific product formation of another cell line are combined by fusing cells. This technology is, for example, the basis of the production of the very valuable and extensive range of monoclonal antibodies.

The issues involved in scaling up animal cell cultures are examined in Chapter 8 which also includes a description of the vessels and bioreactors used for large scale operations. In Chapter 9, the processes and potential applications of embryo and organ cultures are described. The strategies and vectors used to genetically manipulate animal cells in culture are explained in the final chapter.

The strong practical emphasis of this text has been balanced by the inclusion of material that will enable readers to understand the rationale of the techniques that have been described. Although written in the context of the use of animal cell culture in biotechnology, the authors have not limited their horizons to this perspective. The text includes many examples in which the *in vivo* cultivation of animal cells have contributed to our understanding of biological processes. From time to time they touch upon the moral issues raised by these techniques. It was not, however, their intention to deal with the broader legal and ethical issues which arise from the outcomes of animal cell culture studies and products.

The authors are to be congratulated in providing an easy to read, readily understandable description of this important cluster of techniques and procedures.

Scientific and Course Advisors: Professor M C E van Dam-Mieras
Dr C K Leach

Introduction to animal tissue culture

1.1 Introduction   2

1.2 Historical background   3

1.3 The applications of tissue culture   4

1.4 Terminology   8

1.5 Stages in cell culture   11

1.6 Outline of the rest of the text   18

1.7 Animal tissue culture in perspective   18

1.8 Literature   19

Summary and objectives   20

Introduction to animal tissue culture

1.1 Introduction

Every living organism is composed of one or more cells. Our understanding of cells dates back to 1665, when the British scientist, Robert Hooke described certain structures in a piece of cork as *cellulae* which is Latin for 'little rooms'. At about the same time the Dutch scientist, Antonie van Leeuwenhoek, was designing the first simple microscope which proved to be a vital tool in our understanding of cells. Systematic studies of microscopic anatomy were performed by Matthias Schleiden and Theodore Schwann in the 1830s and these led to the formulation of the 'cell theory' which has two basic tenets. Firstly it states that all organisms consist of one or more cells, and secondly that the cell is the basic unit of structure for all organisms. This theory was extended in 1855, when the German physiologist Rudolf Virchow stated that all cells can only arise by cell division from pre-existing cells. Since all life has a cellular basis, much research has been done in order to understand how cells work. In order to benefit from this text you will need to have a fairly good background in cell biology, and if you think this may be a problem, we recommend you work through the BIOTOL text 'Infrastructure and Activity of Cells'.

This chapter will deal with some of the general aspects of *in vitro* cultivation of animal cells, terminology, and assumed knowledge. We will start by looking historically at the development of the methodology. Then we will cover some of the modern applications of the *in vitro* culture of animal cells. We will explain some of the important terms which you will need to know to work through the rest of the text. Finally we will explain the layout of the remainder of the text so that you will see how the story will unfold as you work through it.

1.1.1 Multicellularity and differentiation

One of the characteristics of most animals is that they are multicellular - in other words - they are composed of many cells. With this multicellularity comes the specialisation of cells. In a multicellular organism, each cell does not have to carry out all the activities necessary for the life of the organism. Although most cells of any higher organism have many organelles and metabolic pathways in common, each cell is also unique in expressing some of these components to an enhanced degree in order to fulfil a specific function within the organism. Each cell type has its own role - to secrete a specific product, to contract, to transmit an electrical impulse, and so on. The result of this cellular specialisation is that animals consist of a number of different types of cells - each with a characteristic size, shape, structure and function. Such cells are said to have differentiated. A vertebrate has more than 100 different types of cells. These cells associate in very organised patterns to perform specialised functions.

animals consist of different cell types

Many animal cells can, with special care, be induced to grow outside of their organ or tissue of origin. Isolated cells, tissues or organs can be grown in plastic dishes when they are kept at defined temperatures using an incubator and supplemented with a medium containing cell nutrients and growth factors. The *in vitro* cultivation of organs, tissues and cells is collectively known as tissue culture, and is used in many areas of science.

tissue culture

1.2 Historical background

∏ We will introduce a lot of new terms in the next few pages. Get yourself a piece of paper and construct yourself a glossary as you read. We suggest you use the following headings.

term	explanation or definition

Tissue culture is not a new technique - in the scientific literature there are references to its use dating back to 1885. An embryologist, called Roux, was able to maintain the medullary plate of a chick embryo in warm saline for a few days. This was the first recorded example of successful explantation. In 1903, Jolly made detailed observations on *in vitro* cell survival and cell division using salamander leucocytes. In the early experiments, fragments of tissue were studied, and this gave rise to the name 'tissue culture'. Officially, the term tissue culture is used when cells are maintained *in vitro* for more than 24 hours.

1.2.1 Aseptic technique

aseptic
technique

One of the main difficulties of animals tissue culture was keeping the cells free of contamination. Since bacterial cells grow faster than animal cells, contamination quickly leads to bacterial overgrowth. In 1913, Alexis Carrel applied aseptic techniques to tissue culture. He introduced the 'Carrel flask' (shown in Figure 1.1) which facilitated culture under aseptic conditions.

culture medium

antibiotics

flow cabinets

To meet the nutritional requirements of cells, embryo extracts or animal blood serum were added to the cells. These were particularly vulnerable to contamination but the addition of the antibiotics, penicillin and streptomycin, to the cell culture medium from the 1940s onwards alleviated this problem. The problems of microbial contamination have also been greatly reduced by the use of laminar air flow cabinets, which minimise the possibility of contamination by air-borne microbes.

1.2.2 Cell culture

trypsinisation

cell culture

Another significant advance was the use of trypsin (a proteolytic enzyme) by Rous and Jones in 1916 to free cells from tissue matrix. It was subsequently used for the sub-culture of adherent cells and in the 1950s, the technique of trypsinisation was exploited to produce homogeneous cell strains and this marked the start of animal cell culture techniques. Trypsinisation is the term which applies to the treatment of cells by the proteolytic enzyme trypsin to change their adhesiveness. The term 'cell culture' refers to cultures derived from dispersed cells which have been taken from the original tissue. In cell culture the cells are no longer organised into tissues.

1.2.3 Modern tissue culture

modern tissue
culture

Today, tissue culture techniques are considerably easier than they used to be, with standardised media and sophisticated incubation conditions. In the 1940s and 1950s, tissue culture media were developed and conditions were worked out which closely simulate the situation *in vivo*. In particular, the environment is regulated with regard to the temperature, osmotic pressure, pH, essential metabolites (such as carbohydrates, amino acids, vitamins, proteins and peptides), inorganic ions, hormones and extracellular matrix. Among the biological fluids that proved successful for culturing cells, serum is the most significant. 5-20% serum is usually added to media for optimal cell growth. Serum is an extremely complicated mixture of compounds which includes undefined components, and much work has gone towards creating a chemically defined alternative to serum. We will return to this point in Chapter 3.

1.2.4 Cell types which grow in culture

For many years, most cell types were very difficult to grow in culture, but recent modifications in culture methods have allowed many specialised cells to be grown in culture. Literally thousands of different cell lines have been derived from animal tissues. The list of different cell types which can now be grown in culture includes connective tissue elements such as fibroblasts, skeletal tissue (bone and cartilage), skeletal, cardiac and smooth muscle, epithelial tissue (liver, lung, breast, skin, bladder and kidney), neural cells (glial cells and neurons, although neurons do not proliferate *in vitro*), endocrine cells (adrenal, pituitary, pancreatic islet cells), melanocytes and many different types of tumour cells. The development of these tissue culture techniques owes much to two major branches of medical research: cancer research and virology.

range of cells grown in culture

Cancer research

Human tumour cells were obvious subjects for research, since the discovery by George and Margaret Gey in 1952 that human tumour cells could give rise to continuous cell lines. The first human cell line to be grown continuously in the laboratory was the HeLa cell line, which was obtained from a malignant adenocarcinoma of the uterine cervix. This cell line opened up the possibility of characterising the cells of malignant tumours *in vitro* and is still one of the most popular human cell lines for study. Later, non-malignant rodent cells in culture were used to analyse the effect of chemical carcinogens, viruses and other agents on normal growth. We will expand on this aspect in later sections.

HeLa cell lines

Virology

In 1949, Enders *et al* reported that the poliomyelitis virus could be grown in human embryonic cells in culture which led to the production of the polio vaccine for mass vaccinations in 1954. The first human vaccines were produced in primary monkey kidney cells, but these were later found to carry a virus (called SV40) so they were not considered completely safe for the production of human vaccines. In 1962 a human cell line was developed and used for the production of the vaccine. Between 1954 and 1970 a wide range of human and veterinary vaccines were developed *in vitro* and licensed, including vaccines against measles, rabies, mumps, rubella and foot and mouth disease virus. Standard conditions were needed for the production and assay of viruses, and this led to the commercial supply of reliable media and sera, and greater control of contamination with antibiotics and clean air equipment.

production of vaccines in vitro

1.3 The applications of tissue culture

Π See if you can make a list of the possible uses of animal tissue culture *in vitro*.

We begin the list for you:

- to provide material to study the processes that take place in animal cells;

- to enable the cultivation of viruses to be used as vaccines (see previous section). Usually relatively avirulent (non-pathogenic) strains are used. These, however, when introduced into a body as a vaccine, stimulate the immune system to produce defences (eg antibodies) which will also neutralise related pathogenic strains;

Try to add as many possible applications as you can before you read on.

1.3.1 Monoclonal antibodies

hybridoma
technology

Since the 1950s a range of other products synthesised from animal cells have found commercial application. The next major development was the production of monoclonal antibodies. In 1975 Kohler and Milstein produced the first hybridomas capable of secreting a single, specific antibody. The generation of monoclonal antibodies is indispensable in many areas of biotechnology (see Chapter 7) and is just one example of a routinely used production process which is completely dependent on tissue culture techniques.

1.3.2 Recombinant proteins

Another important development has been the production of recombinant proteins from cultured animal cells. Large scale cultures of animal cells are becoming increasingly important in the production of a range of valuable products. In particular, much effort has been directed to the production of lymphokines, interferons, and hormones like human growth hormone. In 1986, human interferon γ (which is thought to have anti-tumour activity) produced from lymphoblastoid cells in culture was licensed as a therapeutic agent.

1.3.3 Reconstitution and replacement of damaged tissues and cells

tissue grafting

In addition to using cells as factories to make viruses or proteins, there is also great interest in the cultured cells themselves as a product, for use in the reconstruction of damaged tissue or as a replacement of non-functional cells and tissues. Tissue grafting may be necessary when tissue has been lost, or has become dysfunctional for whatever reason. The missing tissue can often be replaced and in some cases even whole organs can be replaced if there is a source of healthy tissue. Unfortunately it is not always easy to find an appropriate source of tissues and organs. One source is from people who have died recently, but the tissue is recognised as foreign by the recipient's immune system, so there are problems with immune rejection which have to be dealt with. Another source is to use tissue from another site in the patient (autografting) and this is possible for skin grafts over relatively small areas, but is not possible for larger burn wounds or for most other types of tissue replacement. Cells cultured *in vitro* would be an obvious potential source of the missing function and this possibility has been examined for a number of diseases. Cultivation of the patient's own cells *in vitro*, in order to generate enough cells and/or to genetically manipulate the cells to replace a missing function, would overcome the problems of immune rejection and of scarce tissue sources.

Living tissue equivalents that survive upon transplantation in animals have been constructed from isolated cells of human skin, thyroid and blood vessels. We will now examine some possibilities for using tissue or cells raised *in vitro* as tissue replacements.

Vascular tissue The latter is particularly important since any reconstituted tissue or organ will only survive and develop if it is in contact with a blood supply. A blood vessel model reconstituted from collagen and cultured vascular cells has been studied in depth with a view to therapeutic tissue reconstruction.

Liver The liver is composed of cells with a remarkable capacity for regeneration and the isolation of lineages of parenchymal cells from the liver holds promise for the correction of organ-specific disease. In animal trials, researchers have shown that injection of new liver cells into a scaffolding of polymer sponge or beads which have been injected into the abdomen can take over some of the functions of a diseased liver prolonging the life of the animals for several months. The whole field of tissue reconstitution is expanding rapidly. We will illustrate this with some examples.

Parkinson's disease and neural grafts Parkinsons's disease is a neurological disorder which causes tremor, rigidity and disturbances of posture. Chemical analysis of brain tissue involved in some types of Parkinsonism has revealed a marked depletion of a neurotransmitter called dopamine and this finding has led to attempts to introduce cells making the missing substances.

Recent developments in the treatment for Parkinson's disease include the replacement of cells in the brain with neural cells from another source by neural transplantation. The beginnings of this technique can be traced back to the last century, although the first neural grafts on humans were not performed until 1982. Studies indicate that these neural grafts are effective because they release certain chemicals which can be utilised by the host brain. There are two sources of tissue which have been used in clinical studies to date. Firstly, human embryonic neural tissue (from aborted human foetuses) has been grafted into the brains of patients with Parkinson's disease and this has met with some success. This approach is complicated by ethical controversy and the possibility of immune rejection of the tissue. Secondly, extracts of the patient's own adrenal medulla have been used in a process known as autografting. The adrenal glands are endocrine glands found just above the kidney and the logic behind this choice of tissue is that the adrenal medulla is derived from the neural crest and could, therefore, be expected to have some of the properties of neural cells. However, autografting of adult tissue has not been very successful.

use of embryonic neural tissue and autografts

One of the aims of researchers now is to use cultured cells to supply the missing chemicals to the brain. This will involve identifying the missing factors which are provided by the neural graft and to find a cell line which makes this factor, or to genetically engineer cultured cells to make the factor. Parkinson's disease is only one example of a disease which may one day be cured using cultured cells.

Immunosuppression therapy Patients may be immunosuppressed for many reasons; for example because their immune systems have been suppressed by the drugs needed to treat cancer or to prevent graft rejection, or by diseases such as AIDS. Such patients become particularly susceptible to infection by pathogens and one of the major threats, striking about 50% of bone marrow transplant recipients is the cytomegalovirus (CMV) which can cause fatal pneumonia. Killer T-cells which specifically attack CMV can be separated from the person donating the bone marrow and large numbers can be grown in culture. In one recent study, such antigen-specific killer T-cells grown *in vitro* were used to inject immunosuppressed patients. Preliminary results indicated that none of the patients suffered from CMV infection after this treatment. One future possibility of this kind of treatment will be to use specific killer T-cells against the HIV virus in order to reduce or prevent the symptoms of AIDS.

1.3.4 Gene targeting

replacement/
correction of
faulty genes

As our understanding of certain genetic disorders increases, the logical direction of these transplantation techniques will be to correct faulty genes before reintroducing the tissue into the patient. Advances have been made in introducing foreign genes by insertional mutagenesis into cultured skin and blood vessel cells. In Chapter 10 we will develop your knowledge of this process and its applications.

1.3.5 Amniocentesis, infertility and embryo transplantation

detection of
genetic
abnormalities
of foetuses

Genetic abnormalities of foetuses may be identified by culturing cells collected from the amnion during early pregnancy. Some of the amniotic fluid is removed by a process called amniocentesis and the cells are cultured to provide enough material for chromosome analysis (Chapter 4). The techniques of animal tissue culture are also directly relevant to *in vitro* fertilisation and embryo transplantation that are employed to circumvent some of the problems of infertility.

1.3.6 Cytotoxicity testing

Tissue culture has been used to screen many anti-cancer drugs since the demonstration in 1950 of clear correlations between the *in vitro* and *in vivo* activities of potential chemotherapeutic agents. Explants of human tumour tissues grown *in vitro* can be tested for chemosensitivity in order to tailor the patient's chemotherapy to suit the individual patient and tumour.

reduction in
problems
associated with
species
differences

At present, evaluation of the effects of carcinogens, toxins and drugs on specific cell types using *in vitro* models supplements the findings found using animal models. The most extensive use of experimental animals is in the safety evaluation of drugs, pesticides, food additives, industrial chemicals and cosmetics which all have to be screened for safety. Safety evaluation of a single drug or other chemical to the stage at which it is marketed may involve the use of as many as 1000 animals. In the UK, several million experimental animals are likely to be used for the purposes of safety evaluation of chemicals even though tissue culture tests would be more suitable both in terms of cost and for ethical reasons. In addition, there is a scientific justification for the tendency towards cytotoxicity testing in tissue culture, namely the realisation that animal models are in many ways inadequate for predicting the effects of chemicals on humans since there are many metabolic differences between species. Cytotoxicity studies involve the analysis of morphological damage or inhibition of zone of outgrowth induced by the chemical being tested.

importance of
tissue culture
to health care
and agriculture

From the range of applications of animal tissue cultures we have described above, you should be impressed by the variety and importance of these applications. Clearly we can identify the potential of using such systems to provide better health care to both humans and animals. But we can also add to this list the use of these techniques in the *in vitro* genetic manipulation of animals to produce animals with desired characteristics or for cloning desired strains of domestic animals. Thus these techniques are important in agriculture as well as in health care.

| SAQ 1.1 | Look through the applications of *in vitro* cultivation of animal cells and see if you can distinguish three categories of product. |

1.4 Terminology

It is important to clarify some of the terms which are used to describe different methods of *in vitro* cultivation of animals cells.

1.4.1 Tissue, organ and cell culture

Tissue culture

Tissue culture is used as a generic term to include the *in vitro* cultivation of organs, tissues and cells. As such, the term is not limited to animal cells, but includes the *in vitro* cultivation of plant cells which is described in the BIOTOL text *'In vitro* Cultivation of Plant Cells', but will not be mentioned further in this text. Tissue culture can be subdivided into two major categories; organ culture and cell culture.

Organ culture

Organ culture refers to a three-dimensional culture of tissue retaining some or all of the histological features of the tissue *in vivo*. The whole organ or part of the organ is maintained in a way that allows differentiation and preservation of architecture, usually by culturing the tissue at the liquid-gas interface on a grid or gel. Organ cultures cannot be propagated and experiments using organ culture generally involve a large degree of experimental variation between replicates, making it difficult to use organ culture for quantitative determinations. We will examine organ culture in more detail in Chapter 9.

Cell culture

Cell culture refers to cultures derived from dispersed cells taken from the original tissue. These cultures have lost their histotypic architecture and often some of the biochemical properties associated with it. However, they can be propagated and hence expanded and divided to give rise to replicate cultures. Cell cultures can be characterised and a defined population can be preserved by freezing.

Π   At a first glance, what specific problems do you think might arise when cells are removed from an organism for further study?

In multicellular organisms cells do not function in isolation, but each cell serves the needs of the whole organism. Hence communication is very important for multicellular existence. *In vivo*, cells in close contact interact with each other, and these interactions are essential in the functioning of tissues and organs. Cells are also in contact with a complex network of secreted proteins and carbohydrates called the extracellular matrix, that fills the spaces between cells. The extracellular matrix helps to bind cells together, provides a lattice through which cells can move, and affects cell behaviour directly. In addition, cells *in vivo* are in contact with hormones and hormone-like factors. This communication between different cell types is lost when cells are cultivated *in vitro*. When cells are grown in culture they are severed from all of the above interactions and can, therefore, not be expected to behave in exactly the way they would *in vivo*. We will come back to this point in Chapter 4.

1.4.2 Adult and embryonic tissue

Cultures can be derived from adult tissue or from embryonic tissue. Cultures derived from embryonic tissue generally survive and grow better than those taken from adult tissue. Tissues from almost all parts of the embryo are easy to culture, whereas tissues from adults are often difficult or impossible to culture. Widely used embryonic cells include mouse embryo fibroblasts 3T3 cell lines and the human foetal lung fibroblast cell lines such as MRC-5 (do not worry about the lettering of these cell lines at the moment. You will learn what some of the important abbreviations mean later). Despite the practical advantages of embryonic cell culture, it is important to remember that these cells may not behave in the same way as adult cells, and they need to be characterised extensively. Chapter 9 will deal with the culture of whole embryos in detail.

1.4.3 Embryonic stem cells

A more recent development has been the removal of embryonic stem cells (ES-cells) from the embryo during the blastocyst stage of development. These cells can be grown in culture for many generations and are of particular interest because they can be manipulated in culture and then re-introduced into embryos.

∏ Write a list of the different categories of cell culture which you now know.

So far we have defined tissue culture according to the origin of the cells (adult or embryonic tissue) and according to the way in which the cells are maintained (organ culture and cell culture). The next divisions we will look at are the different ways in which dispersed cells can grow in culture.

1.4.4 Adherent or suspension cultures

anchorage-dependent cells
contact inhibition

Cells may grow as an adherent monolayer or in suspension. Adherent cells are said to be anchorage-dependent and attachment to a substratum is a prerequisite for proliferation. They are generally subject to contact inhibition, which means they grow as an adherent monolayer and stop dividing when they reach such a density that they touch each other. Most cells, with the exception of mature haemopoietic (haematopoietic) cells and transformed cells, grow in this way.

In contrast to anchorage-dependent cells, cells cultured from blood, spleen or bone marrow adhere poorly if at all to the culture dish. In the body, these cells are held in suspension or are only loosely adherent. It is important to realise this if you are working with this category of cells, since the methods used to propagate these cells are very different to those for adherent cells.

The plasticware which is used for cell culture of both adherent cells, and cells which grow in suspension, is shown in Figure 1.1. Suspension cultures are easier to propagate, since subculture only requires dilution with medium. Cultures in which cells grow attached to each other or to a substratum have to be treated by a protease to break the bond between cells and substratum. The most commonly used enzyme is typsin. Clearly, freely suspended cultures do not require trypsinisation. They are, therefore, also easier to harvest.

∏   Examine Figure 1.1 and see if you can:

1) identify a common feature of the vessels shown.

2) identify a major difference between the vessel shown.

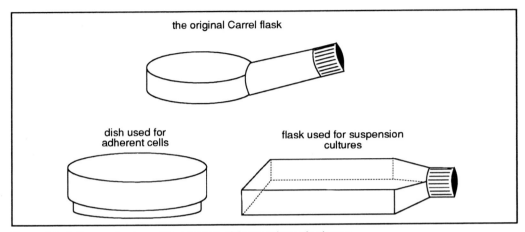

Figure 1.1 A diagram to show different types of tissue culture plasticware.

What we hoped you would identify is the fact that all the vessels shown have a large surface area to which cells may adhere. This is, of course, vital for anchorage-dependent cells if we are to achieve a reasonable cell yield. The main difference is that of the access into the vessel for inoculation and harvesting. Clearly it is easier to inoculate and harvest culture from the dish than from the two flasks shown because by removing the lid we can get to the whole substratum which bears the cells. There is, however, a greater chance of the culture becoming contaminated if it is exposed in this way.

Tumour or normal tissue

Cells derived from normal tissue may not grow indefinitely *in vitro*. This will be covered in more detail in the next section. The only human cells which will grow indefinitely in culture are those obtained from tumour tissue. This means that work with non-malignant human cells needs a constant source of fresh tissue and the problem is that normal human tissue is relatively difficult to obtain. The restriction with respect to the source of tissue means that in practice only a few cell types are commonly used in human tissue culture. Cells which will grow indefinitely *in vitro* are said to be 'immortal'.

immortal cell
lines

SAQ 1.2	Which of the following statements are true and which are false.

1) It should be anticipated that normal (non-malignant) epithelial cells when cultured *in vitro* will grow attached to the surface of the culture vessel.

2) Non-malignant human cells will grow indefinitely *in vitro* and are said to be immortal.

3) Cells derived from embryos are relatively easy to cultivate *in vitro* whereas cells derived from adults are often difficult or impossible to culture.

4) Mature cells derived from haemopoietic (haematopoietic) stem cells show little or no anchorage-dependence.

5) When cultivating cells derived from tumours, it is essential to use vessels with a large surface area to which the cells may adhere.

6) The ability to grow animal cells in culture shows that the extracellular matrix found in natural tissues simply provides a cementing material that binds cells together. |

1.5 Stages in cell culture

1.5.1 Primary cell culture

explants,

primary cell culture

secondary culture

Within the category of cell culture, we can make further subdivisions. We will discuss these in more detail now. The first step in preparing any culture is the sterile dissection of the tissue from the organism concerned. The tissue is chopped into pieces of around 1 mm^3 which are put on a dish. At this stage the tissue is called an explant, which simply means that it is a tissue taken from its original site and transferred to an artificial medium for growth. The cells of the tissue can be isolated and disaggregated by mechanical, chemical or enzymatic digestion of animal tissue. When these cells are induced to grow *in vitro*, a primary cell culture results and these cells are generally still fairly representative of the original tissue. The culture is called a primary culture until it is subcultured for the first time, after which it becomes a secondary culture.

∏ See if you can list two disadvantages you think primary cell culture may have over other types of cell culture.

disadvantages are

Two important disadvantages of primary cell culture over some of the other types of cells culture are that:

heterogeneous cultures

- a mixture of cell types is generally present (since most tissues contain mixtures of cells), in other words it is heterogeneous;

recurrent sacrifice

- primary cell culture requires the recurrent sacrifice of animals.

1.5.2 Passaging cells

lag phase
logarithmic and
stationary
phase

Adherent cells grow as a monolayer until they reach confluence. The kinetics of animal cell growth *in vitro* follows a similar pattern to the classical kinetics demonstrated in cultures of bacteria. When cells are taken (from a tissue, primary culture or stationary phase culture) there is at first a lag phase of some hours or days before the cells begin to grow. Growth then proceeds steadily with the population doubling every 15-20 hours in the case of fast-growing cells. This phase of exponential growth is known as the logarithmic phase. At the end of this phase the maximum population is reached and cells enter the stationary phase or plateau phase in which virtually no growth occurs. At this stage the cells can be trypsinised and reintroduced into fresh media. This is called 'passaging' or 'sub-culturing'. Ideally cells should be passaged during the late log phase of cell growth.

SAQ 1.3

Use an arrow to identify the stage at which cells should ideally be passaged in the graph shown below.

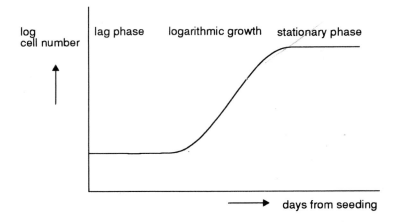

passage
number

The 'passage number' is the number of times this procedure is performed after the original isolation of cells from the primary source. When a primary culture reaches confluence, the cells should be trypsinised and re-seeded in a fresh dish. At this particular stage it is called a secondary culture and more generally from now on through subsequent passaging it is called a cell line. A cell line arises from a primary culture at the time of the first successful subculture and the term implies that cultures consist of lineages of cells originally present in the primary culture. Since the primary culture was heterogeneous, the cell line arising from the primary culture can also be heterogeneous. Selection or cloning of cells with particular properties or markers can be performed on either the primary culture or the cell line, resulting in different cell strains. Thus the term cell strain implies some sort of selection or cloning has taken place. We will return to this point later.

cell line,

cell strains

1.5.3 The fate of primary cultures

continuous - immortalised cell lines finite cell lines

Some cells are capable of an unlimited number of cell divisions *in vitro* as long as they are supplied with nutrients. We can, therefore, sub-culture such cells indefinitely and there is no limit to the number of passages. These are described as continuous or immortalised cell lines or strains. Other cells are only capable of limited numbers of cell divisions after which the culture stops dividing. These are called finite cell lines or strains. The terms finite or continuous are used as prefixes if the status of the culture is known. Immortalised cell lines were previously sometimes called established cell lines. Actually, the term continuous cell lines is somewhat confusing because the term "continuous" in bioprocess technology is usually used to describe the cultivation of cells in a "continuous" rather than a batch system. We would, therefore, prefer to use the term immortalised cell lines to describe cell lines that can be sub-cultured (passaged) indefinitely.

species differences

The replicative capacity of cultured cells varies enormously depending on the cell type and the species. Many cells can be passaged a large number of times, whereas others will die at an early stage. There are some interesting differences between species in their reaction to culturing. In rodent tissue culture, it is not uncommon for cells to divide indefinitely whereas cells of normal human tissue never give rise to continuous cell lines (unless treated with certain agents, which will be discussed in more detail in Chapter 6) and only tumour cells grow indefinitely. Chicken cells are difficult to maintain in culture for any length of time and die after only a few population doublings. The cause for this limited growth potential is not understood, and no nutritional regime has yet been discovered to solve the problem.

In summary, it is difficult to generalise across different species about what happens to cells when they are cultured.

Human cell culture

a finite number of cell divisions

In 1961, Hayflick and Moorhead studied the potential of human foetal lung fibroblasts to divide in culture by counting the cells at each passage starting with the explant from human tissue. They found a slow increase in the growth rate (phase I). During this phase, some cells die, and other cells grow. If the culture is continually diluted they grow at constant rate for an average of 50 generations (phase II) after which the growth rate begins to slow down. The ensuing period of increasing cell death (phase III) ultimately leads to complete death of the culture. Thus, human cells could be 'passaged' or 'sub-cultured' for about 50 population doublings before growth stopped and senescence occurred. Since then, other human cell types have been studied, and the results have generally been similar, although the exact number of doublings depends on the type of cell, its stage of differentiation and origin. The number of population doublings for human cells is generally between 20 and 80 but it can be much shorter. At the end of this stage, the cells first start to look strange and a few weeks later the culture always dies even when apparently provided with all the right nutrients. The limited replicative capacity of human cells in culture is sometimes called the Hayflick effect, after its discoverer.

Hayflick effect

link between the Hayflick effect and aging

Further experiments indicated that the loss of replicative capacity has an interesting link with the process of aging. The average number of population doublings achieved was 50 for cells derived from human foetal lung fibroblasts, but only 20 for cells derived from adult lung fibroblasts. Moreover, experiments with human skin fibroblasts, in 1970, indicated that the number of doublings achieved in culture is a function of the donor's age. A linear decrease in the life span of the cells was found with increasing age of donor. These experiments indicated that there does appear to be some relationship between the aging of cells in culture and the aging of cells in the human body and suggest that cells may have a built in "time clock" which tells them how long to survive.

∏ Starting with 10^5 human diploid cells, how many cells would you have after 50 doublings?

You would have $10^5 \times 2^{50}$ (approximately 10^{20}) cells, thus even though the lifetime of these cells is limited, the culture can still be studied for a long time, and may be a valuable research tool. However, such cultures still have a number of disadvantages. As the cultures gets older, there are changes in the behaviour of the cells so controlled studies are difficult. The finite lineage of cells originating from such primary cultures have different properties at different stages of passaging, so it is important to carefully record the passage number. Also, the limited life-span of these cells means that there is still a need to find fresh tissue when the cells die. The ideal cell culture system would grow indefinitely without variation in cell phenotype.

Rodent cell culture

A fraction of the cell population from a primary culture derived from a rodent may continue to grow beyond the point when cells from many other sources would have stopped growing and died. The culture is designated as continuous or immortal when it has been sub-cultured at least 70 times at an interval of three days between each sub-culture. This often happens in rodent tissue culture experiments and cultures of rodent embryo cells routinely give rise to immortal cultures.

When a culture is prepared from rodent tissue, there is some initial cell death coupled with the emergence of healthy growing cells. As these are diluted and allowed to continue growth, they soon begin to lose growth potential and most of the cells die. The culture at this stage is said to undergo crisis. However, occasionally a variant population of cells emerges from this phase and these cells continue to grow until their progeny overgrow the culture. These cells constitute an immortalised cell line which will grow forever if appropriately diluted and fed with nutrients. A protocol which is much used to establish rodent cell lines is one whereby 3×10^5 cells are transferred every three days to new dishes. The lines made by this protocol are called 3T3 lines. 3T3 fibroblasts are much used in cancer research (see Chapter 6).

crisis

continuous cell
lines

3T3 lines

There is a smooth transition to continuous culture, although the origin of the cells which continue to grow at this stage is not entirely clear. They may result from a mutational event, or alternatively it may be that immortalised cells were already present in the original culture, but that they were masked by the presence of finite populations of cells.

∏ Write down any possible disadvantages that study of continuous cell lines may entail.

One of the main disadvantages of continuous cell lines is that the process of generating the cell line involves selection. Cells which grow well in culture will tend to give rise to cell lines, and these may not be the most representative of cells *in vivo*. Thus, cell lines often have slightly different characteristics from the original cell population. There is often positive selection for faster growing cells. In addition, cell lines often have slightly reduced cell size, higher cloning efficiency, increased tumorigenicity and variable chromosome complement. Cells of continuous cell lines usually have more chromosomes than the cells *in vivo* from which they arose and their chromosome complement may undergo expansion and contraction in culture. The culture is said to be aneuploid (that is having an inappropriate number of chromosomes) and the cells of such a culture are obviously mutants.

aneuploids

In addition to the problem of selection, there are other factors which result in differences between the phenotype of cells *in vivo*, and the phenotype found in cell lines. We mentioned earlier that cells are grown in culture are no longer able to form the cell-cell and cell-extracellular matrix interactions which they would *in vivo* and that these cells usually lack the soluble factors which induce differentiation *in vivo*. This may be the reason for the tendency of cell lines to lose their differentiated properties. This aspect will be discussed in more detail in Chapter 4. For some studies, these characteristics may not be very important, but it is generally important to be aware of these changes.

SAQ 1.4

The graphs in Figure 1.2 show the different phases of cell culture for rodent and human cells. Indicate which is which.

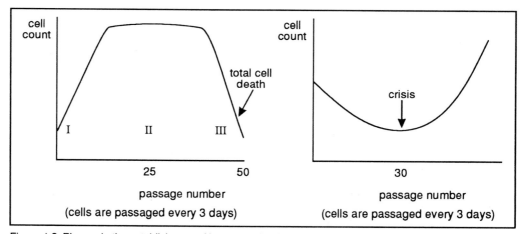

Figure 1.2 Phases in the establishment of human and rodent cell cultures. (see SAQ 1.4).

SAQ 1.5

List three changes which may take place when cells undergo crisis.

1.5.4 Cell cloning

One of the problems mentioned earlier for primary cell culture is that these cultures do not contain only one cell type, but often include a number of different cell types while many biological studies require populations of a single cell type. A number of methods exist to separate different cell types. The traditional approach is to isolate a pure cell strain from cells in continuous culture by a process called cloning. During cloning, a population of cells is derived from a single cell by mitosis to produce a genetically homogeneous clone which can then be characterised and stored. The uniformity of the cells within the cell clone and the potential to increase cell number opens up a wide range of experimental possibilities. A number of different cloning protocols exist based on either physical separation or the use of selective conditions.

cloning

dilution cloning

In dilution cloning, the cells are seeded at low density and are incubated until colonies form as shown in Figure 1.3. They are then isolated and propagated into cell strains.

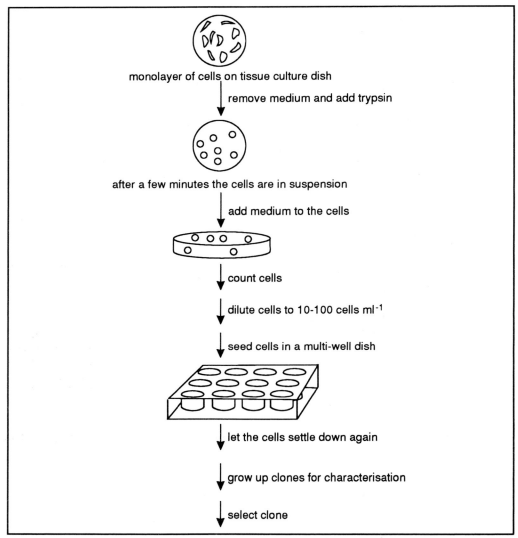

monolayer of cells on tissue culture dish

remove medium and add trypsin

after a few minutes the cells are in suspension

add medium to the cells

count cells

dilute cells to 10-100 cells ml^{-1}

seed cells in a multi-well dish

let the cells settle down again

grow up clones for characterisation

select clone

Figure 1.3 The technique of dilution cloning. The process is designed to deposite a single cell in individual wells in the multi-well dish.

selective media — Another way of isolating cell clones is to use selective media. One of the main problems is that for instance fibroblasts tend to grow well in culture and therefore contaminate other cell types. Fibroblasts can be eliminated either by complement-mediated lysis using monoclonal antibodies against fibroblasts or by using chemicals which suppress fibroblast overgrowth (such as cis-OH-proline, sodium ethylmercurithiosalicylate and phenobarbitone). In complement-mediated lysis, the monoclonal antibodies bind to the fibroblasts. This activates complement (a cascade system of blood proteins) which then lyses those cells to which the antibody is attached.

Note that a clone is a population of cells derived from a single cell and is deemed to consist of a population of genetically identical individuals. Sometimes in the literature the term 'clone' is used to describe populations of cells for which there is no proof that these populations are genetically pure and derived from single cells.

1.5.5 Cell separation

flow cytometry
and flow
cytofluorimetry
FACS

The alternative approach to cell cloning is to separate cells using physical methods. Cells may be selected on the basis of size, density, charge, surface area or specific affinities. Here we will discuss two popular methods; flow cytometry and flow cytofluorimetry measure the light scattering properties of cells, which are proportional to surface area. In flow cytofluorimetry, specific fluorochromes are attached to cells and the cells are separated on the basis of fluorescence emission. Both methods can be performed very quickly using a sophisticated piece of apparatus called the Fluorescence Activated Cell Sorter (FACS). A stream of single cells passes through the flow chamber of the FACS and each particle in suspension is analysed for the appropriate parameter (such as fluorescence emission or light scatter). The cells can then be separated according to these parameters and collected in a sterile manner. Up to 5000 live cells can be sorted in one second using the FACS.

∏ Explain why it might be important to have a homogeneous cell strain.

Scientific experiments must be reproducible and the results must be statistically valid. Therefore, it is important that each replicate culture sample is identical, so that no variation is introduced at this level.

Since there is such a diversity of tissue culture techniques, it is important to consider which technique is the most suitable for a particular study. The primary determinant in selecting a tissue or cell line for study is the nature of the observations to be carried out. General cellular processes, such as analysis of DNA synthesis, can be carried out on almost any cell type whereas more specialised processes can only be carried out on the appropriate cell type, for example analysis of the control of milk production is only feasible using mammary epithelial cells.

SAQ 1.6

Divide the following list of studies into those you would perform using organ culture and those you would perform using cell culture.

Histological studies

Analysis of the mechanism of differentiation

Analysis of cell-cell interactions

Analysis of cell-extracellular matrix interactions

Replicate sampling and quantitation

Propagation of large numbers of identical cells for analysis

Studies on DNA synthesis

Drug toxicity analysis

Production of monoclonal antibodies.

1.6 Outline of the rest of the text

So far, we have looked at the basic principles of tissue culture and mentioned some of the very general problems of different tissue culture methods. In subsequent chapters, we will deal with some of these areas in more depth. In Chapter 2, some of the basic techniques used in tissue culture will be covered in more detail. This chapter will include information on aseptic techniques and safe handling of tissue culture. Chapter 3 will cover the choice of media for use in tissue culture, and new developments in media formulation. We have already briefly mentioned that there are many different types of animal cells performing different functions *in vivo*. In practice, therefore, tissue culture work must involve a decision about which cell type to work with and the media required to cultivate them. Chapter 4 covers the characterisation of different cell types, and analysis of the differentiated state within a specific cell line. In addition, this chapter covers the routine checks which are needed to confirm that no cross-contamination of cell lines or contamination of a cell line with a micro-organisms has taken place. Indefinite maintenance of cell lines in culture is not only impractical but would carry the inherent risks of genetic drift and microbial contamination. Techniques have advanced to permit almost unlimited storage of cell lines at liquid nitrogen temperatures (-196°C). These cryopreservation techniques will be discussed in Chapter 5. Tissue culture can also be used to study the precise controls which govern the rate of cell division. When these controls break down *in vivo*, the result is uncontrolled cell division which is referred to under the general heading of cancer. The use of cultured cells to study the molecular basis of cancer is covered in Chapter 6. Chapter 7 will cover the methodology involved in the production of monoclonal antibodies which are major products of the biotechnology industry.

Another important component of the biotechnology industry is the manufacture of useful proteins from animal cells and Chapter 8 concerns the scale up of animal cell culture to yield specific products on a large scale. Chapter 9 deals with concerns embryo and organ culture. Finally, Chapter 10 covers the genetic manipulation of animal cells in culture.

1.7 Animal tissue culture in perspective

In vitro culture of animal cells and organs enables experiments to be conducted at greater uniformity than can be achieved using whole animals, leading to improved reproducibility between successive experiments. In addition, tissue culture avoids many of the moral and ethical questions of animals experimentation and allows experiments on human tissue which would not be possible *in vivo*. However, it is important to remember that tissue culture models still involve the killing of animals to supply material.

The *in vitro* cultivation of animal cells also has a number of disadvantages over *in vivo* models. Firstly, tissue culture requires a certain amount of skill - culture techniques must be carried out under strict aseptic conditions because animal cells grow less rapidly than many of the common contaminants such as bacteria and yeast. Secondly, the cost of animal tissue culture reagents, and the labour involved in maintaining animal cells in culture, are often greater than the costs involved in most animal work. (This is certainly the case for experiments on rodents, but *in vivo* experiments using dogs, cats or primates are a lot more costly than tissue culture work). Thirdly, cell lines may not reflect the situation *in vivo*, and whatever system is being studied, great care must be taken to analyse the significance of any *in vitro* findings. It is not possible to reproduce the conditions in the living animal exactly. Finally, tissue culture is not always possible, since suitable cell lines may not be available.

Overall, tissue culture and *in vivo* studies complement each other and both have their advantages and disadvantages in specific situations. It is important to consider which technique is most suitable to address any specific question and to use the most appropriate model for the situation. Tissue culture cells do not behave in exactly the same way as cells do *in vivo* but so long as the limitations of the model are appreciated, tissue culture is still a very valuable tool.

1.8 Literature

It is important to emphasise that the *in vitro* cultivation of animal cells is still a developing technology, and although we have tried to give an up-to-date account of the techniques, there are many new developments published every month, so that some of the information may be already out of date by the time you come to read it. There are a number of excellent review journals which you could consult should you want to find out the current status on a particular area of tissue culture. These include the series 'Current Opinions in Biotechnology', 'Trends in Biotechnology' and 'Trends in Cell Biology', which give overviews of the findings in the more specialised journals.

You will also find suggestions for further reading at the end of this text.

Summary and objectives

In this chapter we have attempted to provide you with an overview of the importance of the cultivation of animal tissue *in vitro* and to introduce you to the key terms used in this area of activity. At this stage we have not given much detail of these techniques but indicated how the description of these techniques is arranged in the remainder of the text.

Now that you have completed this chapter you should be able to:

- outline the historical background of *in vitro* cultivation of animals cells;

- explain the meaning of the terms: tissue culture, organ culture, cell culture, primary cultures, cell lines, continuous cell lines;

- describe, in outline, the applications of tissue culture in cancer research, virology and immunology as well as the possible clinical uses;

- show an understanding of the main factors which determine the choice of the tissue culture system used;

- describe the advantages and disadvantages of tissue culture in comparison with *in vivo* systems;

- show an understanding of the roles that tissue culture can play in the laboratory and the limitations of the approach.

Outline of the key techniques of animal cell culture

2.1 Introduction	22
2.2 Setting up the laboratory	22
2.3 Culturing cells	24
2.4 Maintaining the culture	29
2.5 Quantitation of cells in cell culture	31
2.6 Cloning and selecting cell lines	36
2.7 Physical methods of cell separation	38
2.8 Hazards and safety in the cell culture laboratory	42
Summary and objectives	43

Outline of the key techniques of animal cell culture

2.1 Introduction

In this chapter we will outline some of the techniques commonly used in the culture of animal cells. We will also discuss the importance of aseptic techniques and the safety precautions that must be taken to protect the cell culture and the operators. Note that we use the term aseptic techniques to cover the whole range of procedures used to prevent contamination of cultures by unwanted micro-organisms (contaminants). In later chapters, we will discuss aspects of large scale cell culture, characterisation of cell lines and their preservation. The emphasis here is to explain how to establish and maintain cell lines from whole tissue in a simple laboratory setting.

2.2 Setting up the laboratory

Perhaps the biggest pre-occupation of the cell culture biologist is how to prevent contamination of the tissue cultures. The media used to grow the cells also provide excellent nutrition for unwanted organisms. Contamination of cell cultures by bacteria, fungi and mycoplasma often results in the loss of a great deal of time and money. An additional problem, which is discussed in Chapter 4, is the cross contamination of cells from other cell lines. Before we examine how to ensure contamination is kept at a minimum we need to briefly consider how we would detect contamination.

∏ How would you check for contamination of cell lines? We will deal with this in greater detail in Chapter 4. Here we will provide a brief overview.

visual and microscopic observations

Regularly checking aliquots from culture flasks using a microscope can reveal some types of contamination, for example, yeasts and fungi can be seen easily. Bacteria can be Gram stained or plated out on blood agar plates and general cloudiness of medium also indicates yeast or bacterial contamination. Mycoplasmas are more insidious, the medium is not cloudy, the organisms cannot be seen under an ordinary microscope and most mycoplasma species are very difficult to grow on agar plates. Testing for mycoplasma is time consuming but must be done regularly because cross contamination is very easy. DNA stains or molecular probes may be used to detect their presence in cells, those cells which turn out to contain DNA in the cytoplasm must be discarded.

2.2.1 Checking media ingredients and glassware

observation by sub-culture

Much of the contamination comes from the cell culture medium and its components or from inadequately cleaned glassware. Routine sterility checks on the medium, serum and nutrients are recommended. To do this, a small aliquot of the medium should be incubated at room temperature and another similar aliquot simultaneously incubated at 37°C for up to one week before the medium is used for culturing cells. If these small samples are found to be contaminated, the medium should be autoclaved and discarded.

The use of antibiotics in the medium helps to contain the problem to some extent but it should never be seen as an alternative to good aseptic technique and careful monitoring.

2.2.2 Nature of the work area

Where possible, a separate room should be made available for clean cell culture work. This room should be free of through traffic and, if possible, equipped with an air flow cabinet which supplies filtered air around the work surface. A HEPA (High Efficiency Particle Air Filter) filtered air supply is desirable but not always affordable. Primary animal tissue and micro-organisms must not be cultured in or near the cell culture laboratory and the laboratory must be specifically designated for clean cell culture work. Clean laboratory coats should be kept at the entrance and should not be worn outside of this laboratory and brought back in.

All work surface, benches and shelves and the base of the airflow cabinets must be kept clean by frequent swabbing with 70% alcohol or an alternative disinfectant. If an airflow cabinet cannot be provided, the culture work may be done on a clean bench using a bunsen burner to create a sterile 'umbrella' under which the work can be done (see section 2.2.3).

2.2.3 Aseptic techniques

You have probably already some experience of aseptic techniques from culturing micro-organisms. Similar techniques may be used to transfer cultures of animal cells.

Π Write down some basic rules for handling animal cell cultures aseptically and then check with our list below.

The basic rules for aseptic techniques which should be used even if an airflow cabinet is available include:

- if working on the bench, use a bunsen flame to heat the air surrounding the bunsen. This causes the movement of air and contaminants upwards and reduces the chance of contamination entering open vessels. Open all bottles and perform all manoeuvres in this area only;

- swab all bottle tops and necks with 70% alcohol to clean them before opening;

- flame all bottle necks and pipettes by passing very quickly through the hottest part of the flame. This is not necessary with sterile, individually wrapped, plastic flasks and pipettes;

- avoid placing caps and pipettes down on the bench; practice holding bottle tops with the little finger while holding the bottles for pouring or pipetting;

- work either left to right or vice versa, so that all material to be used is on one side and, once finished, is placed on the other side of the bunsen burner. (This may also stop the operator using the same reagent twice!);

- manipulate bottles and flasks carefully. The tops of bottles and flasks must not be touched by the operator. Touching of open vessels should also be prevented when pouring. If necessary practice pouring from one container to another keeping a distance of 5 mm between the two vessels;

- clear up spills immediately and always leave the work area clean and tidy. Dispose of glassware in appropriate bins and discard used plasticware in marked polythene bags for autoclaving or incineration. All glassware or plasticware used for infectious work must always be autoclaved before incineration. Re-usable glassware should be immersed in disinfectant whilst awaiting transfer to an autoclave.

The rules we have listed above are by no means exhaustive. The ways in which they are implemented are, however, slightly different in different laboratories. Learning to apply these rules depends upon gaining 'hands on' experience within a laboratory. You should not attempt to carry out culture transfers without being shown how to do it properly by an experienced operator.

2.2.4 Basic equipment used in cell culture

airflow cabinet easily cleaned benching, incubator with CO_2 supply

In addition to an airflow cabinet and benching which can be easily cleaned, the cell culture laboratory will need to be furnished with an incubator or hot room to maintain the cells at 30-40°C. The incubation temperature will depend on the type of cells being cultivated. Insect cells will grow best at around 30°C while mammalian cells require a temperature of 37°C. It may be necessary to use an incubator which has been designed to allow CO_2 to be supplied from a mains supply or gas cylinder so that an atmosphere of between 2-5% CO_2 is maintained in the incubator.

refrigerator and freezer

A refrigerator or cold room is required to store medium and buffers. A freezer will be needed for keeping pre-aliquoted stocks of serum, nutrients and antibiotics. Reagents may be stored at a temperature of -20°C but if cells are to be preserved it may be necessary to provide liquid nitrogen or a -70°C freezer (see Chapter 5).

microscope(s) waterbaths, centrifuge, counting chambers glass or plastic-ware

A microscope with normal Kohler illumination will be needed for cell counting. An inverted microscope will also be needed for examining flasks and multiwell dishes from underneath. Both microscopes should be equipped with a x10 and a x20 objective and it may be useful to provide a x40 and a x100 objective for the normal microscope. Additional features such as a camera, adaptor and attachments and UV facility may also be required for some purposes. A waterbath and a centrifuge with sealed buckets is also necessary. Additional requirements include counting chambers (Improved Neubauer) or a Coulter counter for counting cells, glass or plastic cell culture flasks, graduated pipettes of various sizes, centrifuge tubes and universal containers, disposable pasteur pipettes, rubber bulbs or automated pipetters for use with pipettes and precise, calibrated pipetters, eg Gilsons, for measuring small volumes from 1-1000 microlitres. More specialised equipment may be needed for particular experiments. Media and supplements are discussed in the next chapter.

2.3 Culturing cells

2.3.1 Sources of tissue for culture

A wide variety of cells from insects, fish, mammals and humans may be grown *in vitro*, some as primary cultures only but many as secondary and continuous cell lines.

Early attempts at culturing tissues relied upon the explantation of whole tissue or organ which could be maintained *in vitro* for only very short periods. Nowadays it is more usual to grow specific cell types from tissues, although there are still some situations where it is necessary to grow a whole organ (or a part of it).

Π   What situations can you envisage where it would be necessary to culture a whole or part of an organ in its intact state?

Some cell functions such as respiration, proliferation and gene transcription can proceed normally even when the cells are isolated from the parent organ. Other functions, such as the production of hormones or a response to external stimuli, are dependent on the interaction of several cell types within the organ.

In organ culture, whole organs or parts of organs in culture are maintained in order to maintain the *in vivo* interactions of the different cell types within the tissue *in vitro*, so that the effect of exogenous stimuli on the whole organ can be studied. For example, pieces of guinea pig or rabbit gut are used to test the effect of atropine-like drugs on the smooth muscle of the gut because nerve synapses and muscle tissue have to work together to produce a measurable response to stimuli.

There are disadvantages to organ cultures. Organs cannot be propagated so each piece of tissue can only be used once, which makes it difficult to assess the reproducibility of a response. And, of course, the particular cells of interest may be very small in number in a given piece of tissue so the response produced may be difficult to detect and quantify. It may not be possible to supply adequate oxygen and nutrients throughout the tissue because of the absence of a functioning vascular system, so necrosis of some cells occurs fairly rapidly. This problem may be ameliorated to some extent by keeping the organ in stirred cultures or in roller bottles which alternately provide air and soluble nutrients. We will provide you with more detail of organ culture in Chapter 9. In this chapter, we will predominantly focus on cell culture techniques.

2.3.2 Sources of cells for culture

Liver, lung, breast, kidney, skin and bladder tissue from animals and humans have been used as sources of epithelial cells for tissue culture. Muscles, bone and cartilage tissue and neural cells can also be grown *in vitro* as can blood cells and many types of tumours cells.

difficulties with adult tissues

Embryo-derived cell lines are easier to establish and sub-culture than tissue derived from newborn or adult animals. With adult tissue it is more difficult to obtain viable proliferating cells because the onset of differentiation is already under way. There is also an increase in fibrous connective tissue which makes disaggregation more difficult and there is a significant reduction in the undifferentiated proliferating cell pool.

Many of the cell lines in common use are widely available from cell banks and type culture collections (see Chapter 5). Some very well known cell lines include the HeLa Line (from a cervical carcinoma) and MRC5 (a fibroblast line developed by MRC laboratories).

Occasionally, however, it is necessary to use primary or low passage cells and these can be prepared by obtaining tissues from laboratory animals or from tissue banks. The whole tissue needs to be disaggregated to produce cells which can be cultured.

2.3.3 Preparation of primary cultures

It is not always necessary to disaggregate tissue before culturing. Some embryonic tissue can be cultured simply by leaving the whole tissue on the flask surface and individual cells will simply grow out from the whole tissue and proliferate. After a few days the original tissue can be removed and the culture medium replenished to allow the new cells to continue growing. This method works for some tissues but it is not suitable for growing specific cells from a piece of tissue.

Tissues that do not require enzymatic disaggregation

∏ Before reading on, see if you can draw a flow scheme of the steps needed to obtain a primary cell culture. Assume that you do not need to disaggregate the cells by enzymatic treatment.

Tissue for primary explant can be treated as follows:

* excess blood is removed by rinsing with a sterile balanced salt solution (BSS). If the tissue is to be transported it can be kept in BSS or medium;

* unwanted material such as fat and cartilage is cut off and the rest is chopped finely using a pair of crossed scalpels. It is important to make clean cuts, the use of tearing actions or scissors may damage the cells;

* the cell suspension is transferred to a sterile 50 ml centrifuge tube together with the buffered saline and the cells are allowed to settle out;

* the medium is carefully pipetted off and the pellet is washed in fresh BSS. The cells are allowed to settle again or the suspension is gently centrifuged at about 1000 rpm for 5 minutes;

* the cell pellet is resuspended in 10-15 ml of medium and the suspension is aliquoted into 2 or 3 25 cm^2 flasks. If the pieces of tissue are still quite large the suspension may be passed through a sieve before culturing, the cultures are incubated at the appropriate temperature for 18-24 hours. (Note that in describing the cultivation of adherent cells, it is usual to describe the size of the vessels in terms of surface area rather than volume);

* if the pieces of tissue have adhered, the medium should be changed weekly until a significant outgrowth of cells has occurred;

* the original tissue pieces which show outgrowth can then be picked off and transferred to a fresh flask;

* the medium in the original flask is replaced and the cells cultivated until they cover at least 50% of the surface available for growth. They can then be sub-cultured if required.

∏ Now that you have read through our scheme, you can either amend your flow diagram or re-draw a flow diagram using the steps we have suggested. You should find this exercise will help you remember the various steps. You may like to use our pictorial representation shown in Figure 2.1 as the basis for your diagram.

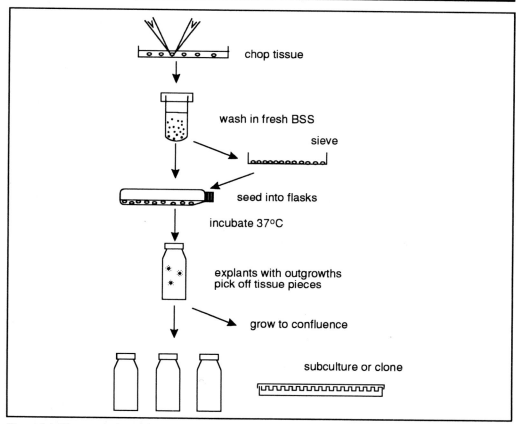

chop tissue

wash in fresh BSS

sieve

seed into flasks

incubate 37°C

explants with outgrowths
pick off tissue pieces

grow to confluence

subculture or clone

Figure 2.1 Disaggregation of tissue and primary culture.

All steps must, of course, be done in such a way as to reduce the chances of the tissue, media and vessels becoming contaminated. Thus all dissection tools, centrifuge tubes and reagents need to be sterilised before use.

This method is useful for small pieces of tissue such as skin biopsies. Fibroblasts, glial cells, epithelium and myoblasts all migrate out of the tissue very successfully. Some selection occurs because not all cells will adhere to the same extent and some cells rapidly outgrow the others.

Tissues that require enzymatic disaggregation

The tissue may be kept whole or may be cut into small pieces as described above. However, densely packed tissue is more difficult to digest and prolonged contact with digestive enzymes such as trypsin often causes destruction of viable cells. Large pieces of tissue for enzymatic disaggregation should be cut into smaller pieces as for the primary explant and digested with trypsin for about 30 minutes at 37°C. Then the trypsin is neutralised by the addition of serum. The amount of trypsin that is required to achieve satisfactory release of cells is tissue-dependent and also depends upon the activity of the enzyme preparation. Trypsin suppliers often provide useful guidance for the use of their own preparations. Typically 0.25% w/v trypsin solution is satisfactory.

use of trypsin

Π We have said that trypsin can be 'neutralised' by the addition of serum. What do
     we mean by this?

Trypsin is a proteolytic enzyme which hydrolyses proteins. By adding a lot of protein
(in the form of serum), the enzyme begins to hydrolyse these proteins rather than the
proteins which bind cells together. What we are really doing is 'diverting' the enzyme
to an alternative substrate so that it is no longer available to attack the extracellular
matrix.

trypsinisation Thus once trypsinisation is complete, medium containing serum should be added to
neutralise enzyme activity. If trypsin is used to disaggregate whole tissue, the
dissociated cells should be harvested after about 30 minutes and washed to remove
trypsin. The remaining whole tissue may be retrypsinised. An alternative method is to
soak the piece of tissue in cold trypsin at 4°C overnight. The cold trypsin penetrates
throughout the whole tissue but will have minimal activity at this temperature. The
tissue can be disaggregated by increasing the temperature to 37°C for 30-40 minutes,
then enzyme activity is neutralised by the additions of medium with serum as before.

collagenase Collagenase and versene (phosphate buffered saline + EDTA) are also used to
and versene disaggregate tissues. Both are gentler in action than trypsin but expensive because
treatments larger amounts are required. Collagenase or versene can also be used together with
trypsin to allow trypsin to be used at lower concentrations.

A variety of other hydrolytic enzymes (for example pronase) have also been employed
to disaggregate tissues in some laboratories.

Once disaggregation is complete, the cells can be washed in fresh BSS or medium and
seeded out in 24 cm^2 flasks. The cell layer can be supplied with fresh medium after 48
hours. This culture is referred to as a primary cell culture.

SAQ 2.1 The enzymes used to disaggregate tissues must be free of microbial
contamination. How would you ensure that this is the case?

SAQ 2.2 Why is collagenase less damaging to cells than trypsin?

2.3.4 Removing non-viable cells from the primary culture

If the primary culture is of anchorage-dependent or adherent cells, the non-viable cells
can be removed by pouring off the medium and rinsing the cell layer with buffer before
adding fresh medium. Cell which are not viable will not be able to adhere to the
substratum.

separation of If the cells are to be cultured in suspension, the non-viable cells gradually become
viable and diluted as the viable cells proliferate. If it is necessary to remove the non-viable cells
dead cells by from the suspension, the cell may be layered onto Ficoll or 'lymphoprep' and be
Ficoll centrifuged at 2000 rpm for 15-20 minutes. The non-viable cells will sink to the bottom
centrifugation and the viable cells can be collected from the medium-Ficoll interface.

2.4 Maintaining the culture

If a primary culture is not to be used as such, it may be sub-cultured to produce a cell line. As we learnt in Chapter 1, cell lines may be of a very limited lifespan or they may be passaged several times before the cells become senescent. Some cells, such as macrophages and neurones do not divide *in vitro* and can only be used as primary cultures.

2.4.1 Sub-culturing from primary to secondary cell culture

sub-culturing
may enable the
production of
cell lines

A primary culture contains a very heterogeneous population of cells from the original explant. Some of these cells will die, some will fail to grow, others will grow quickly and become the dominant cell type present. On sub-culturing the primary cell culture, the dominant types will become even more dominant. We can, therefore, foresee that sub-culturing enables us to produce more homogenous cell populations.

As we explained in Chapter 1, we call the culture produced after sub-culture of a primary culture, a secondary culture. Sub-culturing, therefore, enables us to produce cell lines from our original explant. These cell lines may be further sub-cultured, characterised and cloned.

ΠΙ Suggest some disadvantages of sub-culturing a primary culture to produce cell lines in the manner described above.

Producing a cell line has certain obvious advantages. A homogeneous population of characterised cells can be grown to a large scale, replicates are uniform and this makes designing experiments much easier. There are, however, disadvantages. In establishing a cell line only those cells best suited to the *in vitro* conditions are selected for. These cells may lose some of the differentiated characteristics they had while growing *in vivo* and are prone to genetic instability, particularly if they divide rapidly.

ΠΙ What kinds of strategies may we adopt to produce cell lines of particular cell types?

We really have only two main options here. First, of course, we must choose the appropriate tissue in the primary explant stage. It is no good attempting to produce a particular epithelial cell line if we use tissues which do not contain the appropriate cell type. Secondly, not all cells will grow equally well in the same medium. So, in principle, by selecting our medium carefully we may provide the condition most suited to our cells of interest. We will examine media formulation more thoroughly in the next chapter.

2.4.2 Propagating a cell line

coding of
cultures

Once a cell line is established, it needs to be propagated in order to produce sufficient cells for characterisation and storage, as well as for particular experiments. Details of characterisation and storage are discussed in Chapters 4 and 5.

A cell line is given a name or code which identifies its source (for example HuT, Human T cells) and, if more than one line was developed from the same source, a cell line number is also given (for example HuT 78). If cells from this line are cloned, then a clone number need to be given, eg HuT 78 clone 6D5.

If the cell line is likely to be viable for only a few sub-culturings or generations each generation should be noted, for example HuT 78 clone 6D5/2 or HuT 78 clone 6D5/3.

| SAQ 2.3 | You have received a cell culture which has been given the designation HuB clone 1A2/5. What does this label tell you? |

holding
(maintenance)
medium

Once a culture is confluent (ie the cells cover 60-70% of the growth surface available) it can be transferred into a holding or maintenance medium which provides just enough nutrients to keep the cells alive and healthy but reduces the replication rate so that the cells do not overgrow. The usual method is to use the same basic medium and reduce the amount of serum it contains.

If cells are required for experiments or storage the contents of the flask can be split or divided to seed 2-4 new flasks depending on the vigour of the cell growth.

The content of a flask of cells grown in suspension is very simple to split. This is done in the following way:

- stand the flask upright for 1-5 minutes to allow cells to settle;
- aseptically remove as much medium as possible without disturbing the cells. A 10 ml graduated pipette is best. Discard used medium into a waste pot for autoclaving;
- resuspend cells in the remaining volume, measure and divide between the required number of flasks;
- top up each flask with fresh growth medium and incubate as normal;
- record the cell line code, clone number and the passage number.

Cells which grow as monolayers adhering to the flask surface are a little more difficult to split.

∏ Before reading on, see if you can anticipate the steps that are needed.

The steps are:

- remove medium by pouring it off aseptically;
- rinse the monolayer with prewarmed (37°C) phosphate buffered saline (PBS) or BSS; pour the medium off into discard pot;
- add 10 ml of a 0.25% (w/v) solution of trypsin in saline. Tilt the flask so that all the cells are covered with trypsin. Other enzymes may also be used in the same way;
- pour off the trypsin and incubate the flask at 37°C for 1-10 minutes. Cells which are very susceptible to trypsin should be checked after 1 minute;
- once the cells have begun to round up and are sliding off the flask surface (it may help to tip the side of the flask gently), the trypsin must be neutralised by the addition of 5-10 ml of medium containing serum;

- wash all the cells down from the sides of the flask and aspirate gently using a pipette. This should break up any aggregates and allow for easier cell counting if required. Avoid creating froth as this can lead to contamination and cell damage;

- a small aliquot (0.2 ml) of the cell suspension may be removed for counting if required. Otherwise, measure the total volume of the cell suspension and divide between the required number of flasks and top up and incubate as before;

- remember to record the cell line code, clone number and the number of times it has been divided (the passage number).

So if a flask of fibroblast cells is divided into two new flasks as well as the original flask the following type of labelling system can be used.

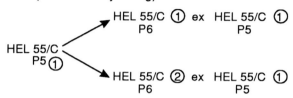

HEL (Human Embryo Lung) cells

Where flasks 1 and 2 are derived from the single flask P5 which represents the passage number.

Note "ex" refers to the flask of origin.

sub-culture at appropriate intervals

The intervals between sub-culturing or changing medium will depend on the cell line and the rate at which it grows. Rapidly dividing cell lines such as HeLa and VERO need to be sub-cultured at least once a week with a medium change in between. Fibroblast HEL cells may be sub-cultured once every 10-14 days. The rate of growth of fibroblasts may be increased, if more frequent sub-culturing is required, by increasing the serum concentration in the medium.

2.5 Quantitation of cells in cell culture

For properly run experiments, it may be necessary to count the cell numbers before, after and even during the experiment. Day to day maintenance of cell lines also requires quantitative assessment of cell growth so that optimum cell densities for sub-culturing and storing can be determined.

We can divide the methods available for determining cell growth into two sub-groups. These are:

- direct methods;

- indirect methods.

In the direct method, cell numbers are determined directly either by counting using a counting chamber or by using an electronic particle counter. In the indirect methods measurement of some parameter such as DNA content or protein content related to cell number is used as the method of estimating biomass. In the following sub-sections, we will discuss each of these approaches in more detail.

2.5.1 Direct methods for quantitation of cells in culture

Counting chambers

The most commonly used device is the Improved Neubauer haemocytometer originally designed for counting blood cells. It consists of a thickened slide with a central chamber of known depth. A grid is etched out (see Figure 2.2) of the silvered chamber bottom.

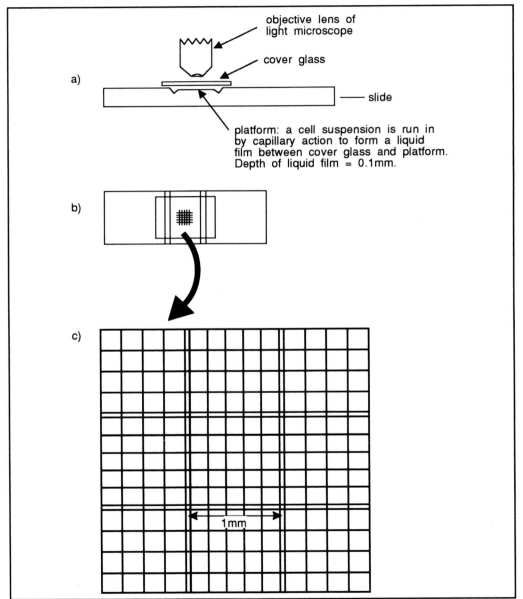

Figure 2.2 A typical counting chamber used to determine cell numbers. Note that different manufacturers use slightly different formats. Some have two sets of grids per slide and the individual blocks of squares are separated by three, rather than two, score lines.

In Figure 2.3 we show how the counting chamber is prepared and loaded with a suspension of single cells for counting. It is important to aspirate the cell suspension adequately before loading the chamber in order to break up clumps of cells which are difficult to count accurately. The counting chamber is examined under a microscope using a x10 objective. The cells in the grid are counted.

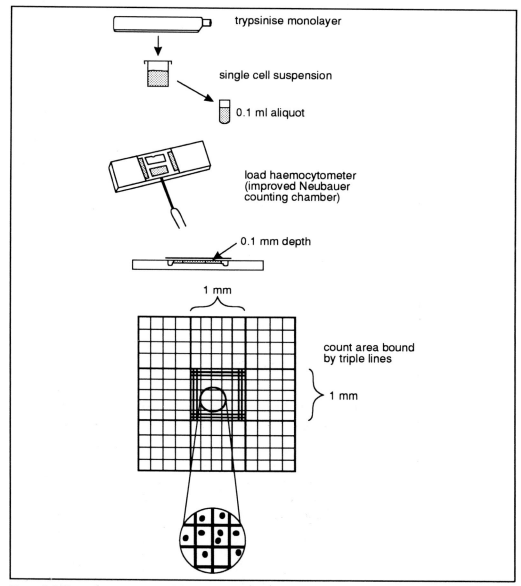

Figure 2.3 Preparing samples and loading haemocytometers for cell counting.

∏ How many cells should we count?

Since the cells are distributed randomly over the grid then, in principle the more cells we count the more accurate our result will be. A good 'rule of thumb' is to use the following relationship. If we count N cells then our error is $\sqrt{N}$. Thus if we count 49 cells then our error will be of the order of 14%, whilst if we count 4900 cells our error is 1.4%. In practice, because of errors elsewhere in the process (for example uncertainties in eluting all cells by trypsinisation), it is pointless attempting to count a very large number of cells. Usually if we count 100 or more cells this will be sufficient for most purposes.

Let us turn this into a practical scheme. First we load the chamber with cells. We can then count all of the cells within the 1 mm square grid. The cell concentration ml^{-1} of suspension can be calculated using the relationship

$$C = \frac{n}{V}$$

where C = cell concentration

n = number of cells counted

V = volume (ml) represented by the grid.

Using the dimensions given in Figure 2.3, the volume of suspension over the grid is: $1 \times 1 \times 0.1 \ mm^3 = 0.1 \ mm^3 = 0.1 \times 10^{-3} \ ml$.

Thus if we counted 100 cells within the grid, the cell density will be:

$$= \frac{100}{0.1 \times 10^{-3}} = 10^6 \ ml^{-1}$$

| SAQ 2.4 |

What would be the cell density, if the number of cells counted was 283?

There are a couple of other practical issues we need to consider.

∏ What would you do if the cell count within the grid was less than 100?

You would need to count the cells in additional grids. Of course, the number of grids counted needs to be taken into consideration when calculating the cell concentration.

∏ Do you count cells which overlap the periphery of the grid?

It is often difficult to decide if cells in contact with the boundary of the grid are more inside or outside of the grid. Thus it is usual to choose two sides (for example the top and the righthand side) before beginning to count. Cells in contact with the boundaries on each of these sides are counted, whilst cells in contact with the boundaries on the remaining two sides are ignored.

0.25% trypan blue dye may be used to check the viability of the cells. The cells are diluted in the dye solution and incubated at room temperature for 1-2 minutes before counting. Cells which take up the stain into the cytoplasm are those which have lost membrane integrity and are non-viable. Healthy, viable cells will exclude the stain.

Coulter counters

Electronic particle counters consist of two electrodes separated by a small orifice:

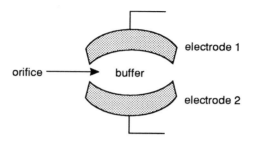

If a potential is applied to the electrodes, current will pass between them through the buffer in the orifice. The amount of current will be dependent upon the conductance (dielectric constant) of the buffer.

∏ What would happen if the buffer contained small particles of a non-conducting substance such as rubber latex?

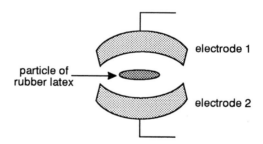

As the particle enters the orifice, the conductance of the solution between the electrodes would be reduced. Thus the current flowing would be reduced and this could be detected electronically.

The size of the change in current flow depends on the size of the particle and the difference in the dielectric constant (conductivity) of the particle and the suspending buffer. This is the principle upon which cell counting using a Coulter counter is carried out.

Cells in suspension are drawn through a fine orifice in the Coulter counter and, as each cell passes through, it produces a change in the current flowing across the orifice. Each change is recorded as a pulse and it is these pulses that are sorted and counted. The size of the pulse is proportional to the volume of the particle passing through, so signals of varying size are produced. The pulse height threshold can be set, therefore, to eliminate electronic noise and weak pulses produced by debris.

Electronic counters, of which the Coulter counter is the most widely used, provide rapid results, but high cell numbers are required to give an accurate count. Another disadvantage is that they cannot distinguish between dead and viable cells and clumps of cells may register as a single pulse thus leading to inaccurate counts.

2.5.2 Indirect methods for determining cells in culture

Other methods of quantitation such as radioisotope labelling and estimation of total DNA or protein are used less frequently. They are useful when cells are grown in microwell plates or as hanging drop cultures. The cells can be mixed *in situ*, stained and counted by eye under a microscope. DNA and protein assays are inaccurate, particularly if cells are multinucleated. They do not distinguish viable and non-viable cells.

2.6 Cloning and selecting cell lines

We have mentioned before that most primary explant cultures consist of very heterogeneous populations of cells, and often when sub-cultured the cells which do not grow under the given conditions are lost. Slow growing cells may rapidly become overgrown by faster growing cells.

Specific cell types can be separated by physical separation techniques or by cloning in order to produce pure cell lines of single cell types. In 1966 Coon and Cahn separated cartilage and pigment-producing cell strains by cloning. The cloned cell lines retained their specialised functions over several generations. In 1978 Clark and Pateman cloned a primary culture of Chinese Hamster Liver to produce a cell line of Kupffer cells.

There are a number of ways to produce a cloned cell line.

These include:

- cloning by dilution;

- cloning by interactions with substrate;

- cloning by selective detachment.

We will discuss these procedures before examining the physical methods available for separating cell lines.

2.6.1 Cloning by dilution

Cells from a heterogeneous culture are trypsinised or agitated to produce a suspension of single cells. The suspension is diluted to give between 10 and 100 cells ml^{-1} by serial dilution; for example a 1 ml aliquot of a suspension containing 1000 cells ml^{-1} is diluted with 9 ml of medium to give 100 cells ml^{-1} and so on. The diluted cell suspension is seeded into microwell plates with the appropriate growth medium and supplements and allowed to grow for 2-3 weeks. The cells from each well are transferred to flasks and grown for characterisation. To some extent specific cell types may also be selected by adjusting the components of the medium to encourage or discourage particular cell types.

Diluted cell suspensions can also be seeded into sterile petri dishes or flasks and grown until small colonies begin to form. The individual colonies can be physically isolated from each other with cloning rings made of porcelain, stainless steel and nylon. The rings, placed around single colonies, simply prevent different colonies growing into one another.

cultures of low cell density may require additional growth factors

It is often necessary to adjust the normal growth medium components when growing clones. Most cells require a minimum number of other cells in their vicinity in order to grow. Cells grown at very low densities may require nutrients from the growth medium which would normally be produced as soluble factors by surrounding cells (for example cytokines such as interleukin 2 which is necessary for T cells to proliferate). These cell-derived products may be present in negligible concentrations in low density cultures.

use of feeder cells

Problems of low density cultures may be overcome by growing the cells in very small volumes eg capillaries or hanging drop cultures. Feeder layers of, for example irradiated mouse cells, can also be used, though the disadvantage with this method is that the cloned cells then need to be separated from the feeder cells. However, irradiated feeder cells do not divide in culture.

conditioned medium

Carefully selected foetal bovine serum, enriched medium or conditioned medium (medium taken from cultures of similar cells which have been grown to 50% confluence, filtered and used to mix with the growth medium for cells to be cloned) may be used to provide essential cell-derived growth factors. The addition of hormones such as hydrocortisone and insulin may also help to improve growth efficiency.

2.6.2 Cloning by interaction with a substratum

separation by adhesion

Different cell types attach to substrata at different rates. This feature can be used to separate different types of cells from a heterogeneous mixture. For instance, if a primary cell suspension is seeded into one flask or petri dish and incubated for 30 minutes and then transferred to a second flask for another 30 minutes and so on, the most adhesive cells will be found in the first flask and the least adhesive ones in the last. Thus macrophages, which adhere very rapidly, will be found mainly in the first flask. Fibroblasts are the next most adhesive followed by epithelial cells. Haemopoetic cells such as T and B cells are the least adhesive and may be found, still in suspension, in the final flasks. In fact, macrophages will often migrate out of whole tissue fragments and attach to the substratum. The rest of the tissue can be lifted off and disaggregated for further separation.

2.6.3 Cloning by selective detachment

use of enzymes and EDTA to selectively detach cells

Some cell types are more easily detached from the substratum by the different mechanisms of action of enzymes such as trypsin and collagenase than are others. Embryonic fibroblasts are readily detached from the flask by a brief exposure to trypsin or collagenase whereas epithelial cells are more sensitive to treatment with the Ca^{2+}-chelator EDTA (Versene).

SAQ 2.5

Below is a scheme for separating specific cell types from a heterogeneous population.

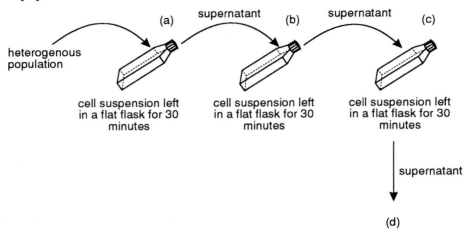

1) If the heterogenous population contains macrophages and T cells, in which vessel or supernatant (a-d) would you anticipate these cells would be found after passage through this process.

2) Are true clones produced by the process described above?

2.7 Physical methods of cell separation

Cloning procedures are time consuming, so by the time a clone has produced enough cells to use, the line may well be close to senescence, particulary if the clones are derived from non-tumour tissues. Tumour tissue and continuous cell lines are cloned more rapidly but, because of the nature of these rapidly dividing cells, considerable heterogeneity can arise even as the clone is being grown up.

Where cells do not grow efficiently enough for cloning, or if time is limited it may be easier to employ physical separation techniques to produce populations of similar cell types. Cells may be separated on the basis of size, density (specific gravity), surface charge or surface chemistry (for example, by the ability to be bound by antibodies or lectins).

2.7.1 Separation based on cell size

The relationship between cell size and sedimentation rate under gravity may be expressed as

$$v = \frac{r^2}{4}$$

where v = sedimentation rate (mm h^{-1}) and r = radius of the cell (μm).

Thus, in principle, cells may be separated on the basis of their size. The relationship shown above is, however, a gross oversimplification and many factors other than the radius of the cell may influence its sedimentation rate.

∏ What other factors influence the rate at which cells sediment?

Cell density may also influence the rate of sedimentation. Obviously the properties of the suspending medium (eg viscosity, density) will also influence the rate of sedimentation. These may be used to improve the separation of different cell types.

2.7.2 Separation based on cell density

For this, a density gradient is established in a centrifuge tube using a suitable density medium. Usually Percoll, Metrazamide or Ficoll is used for this purpose.

isopycnic
sedimentation

The cells are layered on the surface of the density gradient and centrifuged so that the cells sediment to a point where the density is equivalent to their own density. The cell layer can then be siphoned off with a pasteur pipette. This technique is known as isopycnic sedimentation.

2.7.3 Separation based on cell surface charge

With this technique the cells are separated by electrophoresis. The cell suspension is passed between two electrodes (polarised plates) and the cells migrate to either plate according to their net charge ie cells bearing a net negative charge migrate towards the positive plate and vice versa.

∏ Can you identify a major problem with using electrophoresis to separate cells? (Think about the consequence of placing the electrodes horizontally or vertically).

The real problem with this procedure is gravity. If the electrodes are placed vertically, all the cells will tend to sediment onto the bottom electrode. Thus:

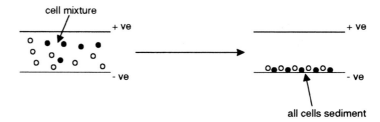

If we try to overcome this by using a higher voltage, there is a tendency for localised overheating and, hence, damage to the cells.

If the electrodes are held vertically, we will still tend to get cells settling out.

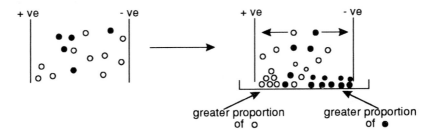

However, some separation of cells can be achieved. (Note that urokinase producing cells were separated from a heterogenous kidney cell population using electrophoresis aboard an American satellite. The zero gravity enabled better separation of cells. This approach is not open to all!!)

A further difficulty is that populations of cells do not usually show qualitative differences in surface charge and we are usually faced with trying to separate cells which are all negatively charged. In these cases, we are dependent upon separating cells on the basis of their net charge:mass ratios. Despite these difficulties, electrophoresis is the basis of cell separation used in cytofluorometry (see section 2.7.5).

2.7.4 Separation based on affinity

antibodies and lectins

Cells in suspension can be passed through affinity columns which contain a matrix coated with antibodies or lectins. As the cells pass through they bind to or are captured by the matrix and can be specifically eluted by washing the column with detergent or enzyme solutions. Care has to be taken not to damage the cells during this process.

2.7.5 Separation by cytofluorometry

FACS

Fluorescent dyes called fluorochromes can be used to label cell surfaces or cytoplasmic molecules. The cells are separated by passing the cell mixture through the flow chamber of a machine such as the Fluorescence Activated Cell Sorter FACS (Becton-Dickinson) or a cytofluorograph (Ortho) or a Coulter Cell Sorter. A Fluorescence Activated Cell Sorter is illustrated in Figure 2.4.

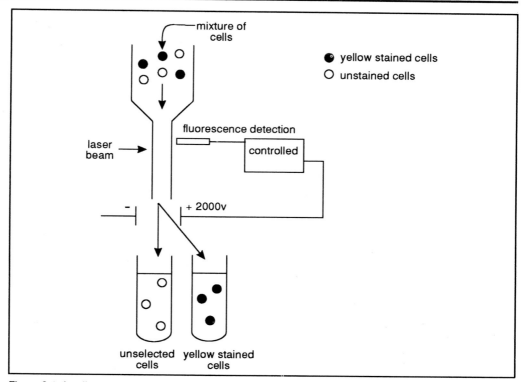

Figure 2.4 A cell sorter based on cytofluorometry. (See text for description).

The fluorescent light produced by the cells under ultra violet light is detected by a photomultiplier and recorded. The cell sorter diverts cells which emit light within pre-set emission boundaries into a receiver tube placed below the cell strains. Those cells which do not emit within the pre-set emission boundaries are diverted to another tube. Consider the yellow stained cells in Figure 2.4. If one of these cells pass through the laser beam, it fluoresces and this is detected by the fluorescence detector. This, through the controller, switches on the voltage at the plates. The negatively charged cell is diverted into the right hand tube. After a short time (that is after the passage of the cell), the voltage is switched off.

If, on the other hand, an unstained cell passes into the laster beam, there is no fluorescence and the voltage is not switched on. Thus, such a cell is not diverted and passes straight through and enters the left hand tube.

This process is very fast and many cells can be separated in a matter of seconds.

This method can be used to separate cells on the basis of any differences that can be tagged by fluorescent labels, eg DNA, RNA, protein, enzyme activity or specific antigens or immunological markers on the cell surface such as HLA or viral antigens. The disadvantage of this method is that the cell yield is very limited and the equipment is very expensive both to buy and to run.

2.7.6 Separation by flow cytometry

This works in a similar way to separation by cytofluoremetry except that the light scattering potential of cells rather than fluorescent labels as used to distinguish between cell types.

2.8 Hazards and safety in the cell culture laboratory

∏ Suggest some biological and chemical hazards you are likely to encounter in a laboratory using tissue culture.

carcinogens and other hazards

Many of the reagents used in laboratories are hazardous if ingested or absorbed through the skin. Chemicals such as DMSO, dyes for staining cells and mutagens for transforming cells are often carcinogenic or mutagenic and must be handled with care. Gloves and laboratory coats must always be worn, masks should be worn when weighing out powdered dyes.

viruses, prions, oncogenes

Biohazards, in the form of viruses, prions and oncogenes, must also be taken into consideration when working with animal tissue. Blood and blood products may contain Hepatitis viruses and HIV, brain and CNS tissue may carry prions which can cause Creutzfield Jacob disease. Animal tissues, particularly from primates may contain Herpes B viruses which are lethal to humans though often carried asymptomatically by the natural hosts.

If tumour cells gain access to the body, they may continue to grow *in vivo* and produce a tumour in the new host.

We will discuss various aspects of safety in later chapters.

Summary and objectives

In this chapter we have looked at some of the core techniques used in the culture of animal cells. We have discussed some sources of tissue for preparing cell lines, the maintenance of cell lines and some methods for measuring cell growth. We have also explained how specific cell types may be selected from mixed cell suspension.

Now that you have completed this chapter you should be able to:

- explain how to obtain and prepare cells for primary cultures;

- describe how to sub-culture primary cell cultures to obtain secondary cultures and cell lines;

- describe a range of direct and indirect methods for determining biomass (cell numbers);

- calculate biomass (cell numbers) using data from haemocytometers;

- describe and compare the different methods of cloning and separating cells;

- discuss the sources of contamination and the importance of aseptic techniques;

- show awareness of the need to take adequate safety precautions against chemical and biological hazards in the laboratory.

Chapter 2

Animal cell culture media

3.1 Introduction   46

3.2 General considerations in media design   46

3.3 Natural media   48

3.4 Synthetic media   49

3.5 Further considerations in media formulation   63

3.6 Nutritional components of media   66

3.7 The role of serum in cell culture   67

3.8 Choosing a medium for different cell types   70

Summary and objectives   72

Animal cell culture media

3.1 Introduction

Explant tissue and the resulting cell lines need to be maintained in an environment which provides the optimum conditions for growth and survival. The cell culture medium is probably the most important single factor in promoting cell survival and proliferation. The medium must provide all the essential nutrition, buffering and gas exchange the tissue needs.

As you might have anticipated, many different media formulations have been developed to meet the requirements of particular cell types or to achieve particular objectives. Broadly we can classify media into four groups:

- media designed to achieve immediate, short term survival of cells;

- media designed to achieve prolonged survival of cells;

- media designed to achieve long term growth of cells;

- media designed to achieve specialised objectives.

In this chapter, we will predominantly focus attention on media designed for short term survival of cells and on media used to achieve prolonged survival and growth of cells.

Our approach here is not to provide simply a catalogue of recipes that have been used. Instead we aim to alert you to the main considerations that underpin the formulation or selection of appropriate media to achieve particular objectives. We will, however, include some specific examples.

We will begin with a few general considerations before describing the range of 'natural' media that may be used. The bulk of the chapter, however, considers synthetic media. In this, we will first discuss so called balanced salt solutions and then describe partially complete and complete synthetic media. Having introduced you to the range of media that have been developed, we will examine a variety of factors that need to be considered in the preparation and use of such media. In addition to this we will consider the roles that the various components play and consider the advantages and disadvantages of using serum to supplement media. In the final part of the chapter, we will briefly consider the selection and design of media to achieve satisfactory performance of different cell types.

3.2 General considerations in media design

In most instances we need to make an early decision as to whether we wish simply to enable cells to survive in culture or to grow and divide.

∏ What do you think will be the fundamental difference in these two types of media?

salts, glucose

Immediate survival media simply need to provide a source of energy and to maintain the correct osmolarity. These basic requirements may be met by a combination of inorganic salts and glucose. Such conditions can be achieved by a defined medium consisting of a balanced salt solution containing glucose.

In contrast, culture media for long term cultivation of cells need to be more sophisticated with a variety of factors.

∏ Make a list of what these factors might be.

vitamins,
amino acids,
growth factors,
hormones

Your list should have included compounds which the cells cannot make for themselves. These might include a variety of amino acids and vitamins. The exact requirements will, of course, depend on the particular cell type. You should have also included specific growth factors and/or hormones since the proliferation of animal cells is usually regulated by such factors. These requirements may be fulfilled in one or two basic ways. First, each of the required ingredients may be supplied as a pure chemical added in controlled (weighed) amounts. Such a method of media preparation would lead to the production of synthetic media. This type of media could also be described as defined media as we know exactly what is present.

An alternative approach to the supply of essential ingredients is to use a natural mixture. Usually this is achieved by adding serum to the media. Since we do not know exactly the chemical formulation of serum, such media are undefined media.

∏ So far we have focused on the chemical composition of the media. Make a list of the other factors which need to be considered.

pH, osmotic
pressure,
surface tension
viscosity

Media not only need to supply the chemicals needed for growth but they must also provide the right physical conditions for cells. For example, the media must maintain an appropriate pH and osmotic pressure. As we shall see later, the surface tension and viscosity of the media are also important considerations.

Finally in this section, we need to mention that it is necessary to consider such factors as the solubility of materials, the compatibility of media components, the purity of material, the stability of the chemicals used and the ability to produce sterilisable mixtures.

Let us illustrate this with a few examples. Salts of Ca^{2+} and Mg^{2+} are generally water soluble. However, in the presence of high levels of phosphate, insoluble calcium and magnesium phosphates are produced. Thus, when phosphates are used in high concentration to buffer a medium, we are limited in the concentration of Ca^{2+} and Mg^{2+} we can include in the media. Some vitamins and amino acids are relatively chemically unstable and thus care has to be taken to ensure that their concentrations remain appropriate. Many lipid components are only sparingly soluble in water and thus care has to be taken to achieve appropriate concentrations.

In this section, we have distinguished two types of media, synthetic and natural. In practice very often a combination of these two types is used. For example, media may be composed of a combination of salts and carbon compounds (to act as an energy source or to fulfil specific requirements) supplemented by serum. Such media are sometimes referred to as semi-synthetic media.

SAQ 3.1

1) Is a medium containing weighed amounts of mineral salts, glucose and a variety of vitamins and organic acids together with a portion of serum a defined or undefined medium? (Give reasons).

2) In the formulation of a medium, high concentrations of calcium and magnesium are to be used. Suggest an alternative buffer to phosphate which could be used to maintain the pH at about 7.4.

3) In the medium described in 2, can the use of phosphates be completely excluded if the medium is to be used to cultivate cells?

3.3 Natural media

A variety of naturally occurring materials are used to cultivate animal cells. These include:

- coagula (for example plasma clots);

- biological fluids (for example serum, tissue extracts and, especially, embryo extracts).

We will not go into the details of how these materials are prepared since, for most purposes, these are obtained from commercial suppliers. We will, however, briefly mention the range of materials used.

3.3.1 Plasma clots

Plasma is rarely used for modern cell culture. Traditionally it was used for growing small pieces of tissue and was particularly useful for avian tissue culture. Because of the limited use of this material we will not elaborate on it further here.

If you wish to find out more about plasma for cell cultivation, the catalogues of media suppliers often provide useful information.

3.3.2 Biological fluids

By far the most commonly used biological fluid is serum. The preparation of serum is relatively straight forward. Whole blood, without anticoagulants, clots rapidly and serum can be simply pipetted from the top of the clot. Serum is usually filtered before use to ensure sterility.

∏ How would you check for sterility of a serum preparation?

Usually this involves inoculating a variety of rich microbiological media with portions of the serum and incubating at 30-37°C for several days. The inoculated media are inspected at specific time intervals over several days. Such tests reveal bacterial contamination. More specific tests are needed for mycoplasma and viruses. These include cultivation under specific conditions or immunological assays. We will discuss these sterility tests in more detail in later sections.

Most researchers obtain the serum they use for the cultivation of cells from commercial suppliers. Other biological fluids that are used in some circumstances include:

- amniotic fluid;

- ascitic and pleural fluids;

- aqueous humour;

- insect haemolymph;

- coconut milk;

- chick (and other embryo) extracts.

Again most of these are available commercially. Their uses are, however, rather limited.

We must alert you to the different forms of serum that are available. Serum may be supplied as whole serum, serum ultrafiltrates or dialysed serum.

Π   What are the main differences between these various forms of serum?

Ultrafiltration removes large molecules (proteins and protein-bound materials) from the serum. In many instances, some of these large molecular weight components may be essential to cells and therefore, serum ultrafiltrates will not support their growth. The main use of serum ultrafiltrates is to provide a supplementary source of nutrients.

When serum is dialysed the large molecular weight components are retained, but the low molecular weight components are lost. Dialysed serum is usually used in semi-synthetic media in which the low molecular weight components are defined. In these cases, the serum does not contribute to the low molecular weight components but supplies the essential large molecules.

SAQ 3.2	Do you think that all batches of commercially available serum will support the growth of a particular cell line to exactly the same extent? (Give reasons for your answer).

3.4 Synthetic media

Synthetic media contains a variety of mineral salts. These are usually described as balanced salt solutions. In this section we will begin by describing such salt solutions before going on to describe more complex media.

3.4.1 Balanced salt solutions (BSS)

These are simple synthetic media which are designed to maintain pH and osmotic pressure as well as provide adequate concentrations of essential inorganic ions. There are many different formulations and some examples are illustrated in Table 3.1.

Substance	Ringer g l^{-1}	Tyrode g l^{-1}	Simms g l^{-1}	Earle g l^{-1}	Hanks g l^{-1}	Dulbecco* (PBS) g l^{-1}
NaCl	9.00	8.00	8.00	6.80	8.00	8.00
KCl	0.42	0.20	0.20	0.40	0.40	0.20
CaCl$_2$	0.25	0.20	0.147	0.20	0.14	0.10
MgSO$_4$.7H$_2$O				0.10	0.10	
MgCl$_2$.6H$_2$O		0.10	0.20		0.10	0.10
NaH$_2$PO$_4$.H$_2$O		0.05		0.125		
Na$_2$HPO$_4$.2H$_2$O			0.21		0.06	1.15
KH$_2$PO$_4$					0.06	0.2
Glucose		1.00	1.00	1.00	1.00	
Phenol Red			0.05	0.05	0.02	
NaHCO$_3$		1.00	1.00	2.20	0.35	
Gas phase	air	air	2% CO$_2$ in air	5% CO$_2$ in air	air	air

Table 3.1 Examples of balanced salt solutions used for mammalian and avian cells. * PBS = phosphate buffered saline

We can distinguish two main types of BSS, those that are designed for use with cultures incubated with air and those designed to equilibrate with high concentrations of CO$_2$. Commonly media based on those of Earle and Hanks are used for these two different conditions.

☐ What is the principle difference between media used with air and those used with air supplemented with CO$_2$?

There are many differences but the main one is that those media to be used with air supplemented with CO$_2$ invariably have high concentrations of bicarbonate (HCO$_3^-$).

Phosphate buffered saline (Dulbecco medium or PBS) is widely used by virologists. Phosphates are relatively weak buffers and not at their most effective at physiological pH and temperature. The pH needs to be adjusted with NaOH and glucose must be added. Dulbecco's PBS buffers are commercially available as tablets which can be dissolved in sterile water to produce a solution at the correct pH without adjustment.

Phosphates or bicarbonates are usually used for buffering media and balanced salt solutions, but tris (tris (hydroxymethyl) aminomethane) and sodium phosphate have also been used. Cell lines such as L, HeLa, HLM and BHK21 can be kept proliferating for several years in media containing tris-citrate.

A wide variety of buffers are now available which have pKa values around pH 6.8-pH 8.0. Many of these are suitable for use in tissue culture media. Many of these are known by their acronyms which are much easier to remember than their full names. Examples are:

- MOPS (pKa 7.2);

- TES (pKa 7.4);

- HEPES (pKa 7.6);

- DIPSO (pKa 7.6);

- HEPPSO (pKa 7.8).

HEPES is most popular

The buffering agent HEPES [(N-2-hydroxyethyl piperazine-N)] ethansulphonic acid is the most popular. HEPES is commonly added to defined and synthetic media but is not used with CO_2 enriched atmospheres.

The main decision in selecting a BSS, therefore, depends upon whether or not the cells are to be maintained under high CO_2 atmospheres and on the nature of the buffer. For the most part, the concentrations of salts used are rather similar in all media. Thus all BBSs contain fairly high concentrations of NaCl and contain K^+, Ca^{2+}, Mg^{2+} and HPO_4^{2-} and $H_2PO_4^-$. The inclusion of other buffers such as HEPES enables us to keep the level of phosphate to a minimum.

∏ What is the role of phenol red in some BSSs?

It acts as a pH indicator.

SAQ 3.3

Describe the potential advantages and disadvantages of using phenol red as a pH indicator rather than using a pH probe.

Before we consider the more complex media, we will elaborate a little on some of the practical issues involved in the preparation of BSSs.

∏ From what we described in section 3.2, what is the most likely problem we will encounter in preparing media of the type described in Table 3.1?

The chemicals described in this table are reasonably stable. The main problem will be that the cations Ca^{2+} and Mg^{2+} and the anions such as PO_4^{3-}, HPO_4^{2-}, $H_2PO_4^-$ and CO_3^{2-} tend to interact and precipitate.

To avoid this the calcium and magnesium salts are usually weighed and dissolved separately into a large volume of liquid. The Ca^{2+} and Mg^{2+} solutions are then added slowly with vigorous mixing to the large volume salt solution. Finally the volume is adjusted using water to produce the final solution.

It is usual to prepare concentrated stock solutions of salt mixtures (say x 10 the normal concentration they are used at) and to store them in polythene bottles. Aliquots of these are added to sterile water to produce the appropriate final concentration.

∏ Suggest some reasons why this can be advantageous and also suggest some potential problems which may arise from this procedure.

The main advantages is that large quantities of media may be brought or made up and stored without taking up much refrigerator space, an important concideration when batches of media are likely to vary. The main disadvantage is that dilution of the stocks to produce working concentrations must be done under aseptic conditions and using good quality, sterilised deionised water. Sterility needs to be checked after dilution and this process can be time consuming.

3.4.2 Partially complete (semi-synthetic) and complete synthetic media

medium 199 or
Morgan,
Morton and
Parker

Partially complete synthetic media were first studied systematically by Albert Fischer (1941) who used dialysed plasma as a basal medium. Later Morgan, Morton and Parker (1950) developed medium 199 (see Table 3.2) which is a complex medium which is still widely used in the production of viruses and vaccines.

∏ Examine Table 3.2 carefully. You should be able to identify components which are constituents of proteins, those which act as precursors of RNA and DNA and a variety of vitamins especially those which function as co-enzymes in enzyme catalysed reactions.

Obviously this is quite a complex medium. Even so it is unsatisfactory for many cell types. Generally, synthetic media without serum supplements can only keep cells alive for 1-2 days. In practice, the addition of serum can help to maintain many cell lines for prolonged periods.

Compound	Concentration (mg l⁻¹)	Compound	Concentration (mg l⁻¹)
Amino Acids		**Vitamins**	
L-Arginine	70.0	Thiamin	0.01
L-Histidine	20.0	Riboflavin	0.010
L-Lysine	70.0	Pyridoxine	0.025
L-Tyrosine	40.0	Pyridoxal	0.025
DL-Tryptophan	20.0	Niacin	0.025
DL-Phenylalanine	50.0	Niacinamide	0.025
L-Cystine	20.0	Pantothenate	0.01
DL-Methionine	30.0	Biotin	0.01
DL-Serine	50.0	Folic acid	0.01
DL-Threonine	60.0	Choline	0.50
DL-Leucine	120.0	Inositol	0.05
DL-Isoleucine	40.0	p-Aminobenzoic acid	0.05
DL-Valine	50.0	Vitamin A	0.10
DL-Glutamic acid	150.0	Calciferol (Vit D)	0.10
DL-Aspartic acid	60.0	Menadione (Vit K)	0.01
DL-Alanine	50.0	α-Tocopherol phosphate (Vit E)	0.01
L-Proline	40.0	Ascorbic acid	0.05
L-Hydroxyproline	10.0		
Glycine	50.0	**Miscellaneous**	
Cysteine	0.1	Glutathione	0.05
		Cholesterol	0.2
Nucleic Acid Bases		Tween 80 (oleic acid)	20.0
Adenine	10.0	Sodium acetate	50.0
Guanine	0.3	L-Glutamine	100.0
Xanthine	0.3	Adenosine triphosphate	10.0
Hypoxanthine	0.3	Adenylic acid	0.2
Thymine	0.3	Ferric nitrate	0.1
Uracil	0.3	Ribose	0.5
		Deoxyribose	0.5

Table 3.2 Morgan, Morton and Parker's medium No 199 (1950). (Data from Paul J. Cell and Tissue Culture, Churchill Livingston, Endinburgh, 1975)

Parker and Healy Medium 858, and NCTC 109 medium

After the introduction of medium 199 many other media of increasing complexity were developed. Medium 858 of Parker and Healy and NCTC 109 (see Tables 3.3 and 3.4) were the first media to be developed which could support cells without the addition of any serum. It is quite likely, however, that the cells which were kept alive in this type of medium were not normal and had become adapted to grow in these conditions.

Compound	Concentration (mg l^{-1})	Compound	Concentration (mg l^{-1})
Amino Acids		**Lipid Sources**	
L-Arginine	70.0	Tween 80 (oleic acid)	5.0
L-Histidine	20.0	Cholesterol	0.2
L-Lysine	70.0		
L-Tyrosine	40.0	**Nucleic Acid Derivatives**	
L-Tryptophan	20.0	Adenine deoxyriboside	10.0
L-Phenylalanine	50.0	Guanine deoxyriboside	10.0
L-Cystine	20.0	Cytosine deoxyriboside	10.0
L-Methionine	30.0	5-Methylcytidine	0.1
L-Serine	50.0	Thymidine	10.0
L-Threonine	60.0		
L-Leucine	120.0		
L-Isoleucine	40.0	**Miscellaneous**	
L-Valine	50.0	Sodium acetate	50.0
L-Glutamic acid	150.0	D-Gluatmine	3.6
L-Aspartic acid	60.0	L-Glutamine	100.0
L-Alanine	50.0	D-Glucose	1000.0
L-Proline	40.0	Phenol red (pH indicator)	20.0
L-Hydroxyproline	10.0	Ethanol (as an initial solvent for fat-soluble constituents)	16.0
Glycine	50.0		
L-Cysteine	260.0	**Antibiotics**	
Vitamins		Sodium penicillin G (added just before use)	1.0
Pyridoxine	0.025	Dihydrostreptomycine sulphate	100.0
Pyridoxal	0.025	n-Butyl parahydroxybenzoate	0.2
Biotin	0.01		
Folic acid	0.01	**Inorganic Salts**	
Choline	0.50	NaCl	6800.0
Inositol	0.05	KCl	400.0
p-Aminobenzoic acid	0.05	CaCl$_2$	200.0
Vitamin A	0.10	MgSO$_4$.7H$_2$O	200.0
Ascorbic acid (Vit C)	50.00	NaH$_2$PO$_4$.H$_2$O	140.0
Calciferol (Vit D)	0.10	NaHCO$_3$	2200.0
α-Tocopherol phosphate (Vit E)	0.01	Fe, as Fe(NO$_3$)$_3$	0.1
Menadione (Vit K)	0.01		

Table 3.3 Synthetic medium No 858 (Parker and Healy 1955).

Co-enzymes			
NAD (95% pure)	7.0	FAD (60% pure)	1.0
NADP (80% pure)	1.0	UTP (90% pure)	1.0
CoA (75% pure)	2.5	Glutathione (100% pure)	10.0
Thiamine pyrophosphate (88% pure)	1.0		

Table 3.3 Continued. (Data from Paul J. Cell and Tissue Cultures, Churchill Livingstone, Edinburgh, 1975).

Compound	Concentration (mg l⁻¹)	Compound	Concentration (mg l⁻¹)
Amino Acids		**Cofactors**	
L-Alanine	31.48	NAD	7.0
L-Alpha amino butyric acid	5.51	NADP sodium salt	1.0
L-Arginine	25.76	Co-enzyme A	2.5
L-Asparagine	8.09	Thiamine pyrophosphate	1.0
L-Aspartic acid	9.91	FAD	1.0
L-Cystine	10.49		
L-Cysteine	26.00	**Nucleic Acid Derivatives**	
D-Glucosamine	3.20	UTP	1.0
L-Glutamic acid	8.26	Deoxyadenoisine	10.0
L-Glutamine	135.73	Deoxycytidine-HCl	10.0
Glycine	13.51	Deoxyguanosine	10.0
L-Histidine	19.73	Thymidine	10.0
Hydroxy-L-Proline	4.09	5-methyl-cytosine	0.1
L-Isoleucine	18.04		
L-Leucine	20.44	**Miscellaneous**	
L-Lysine	30.75	Tween 80	1.25
L-Methionine	4.44	Glucuronolactone	1.8
L-Ornithine	7.38	Sodium glucuronate	1.8
L-Phenylalanine	16.53	Sodium acetate	50.0
L-Proline	6.13	Phenol red	20.0
L-Serine	10.75	Sodium Chloride	6800.0
L-Taurine	4.188	Potassium Chloride	400.0
L-Threonine	18.93	Calcium Chloride	200.0
L-Tryptophan	17.50	Magnesium sulphate	200.0
L-Tyrosine	16.44	Sodium monobasic phosphate	140.0
L-Valine	25.00	Sodium bicarbonate	2200.0
		Glucose	1000.0
		Anti-oxidants	
		Glutathione - monosodium salt	10.10
		Ascorbic acid	49.90
		Cysteine hydrochloride	259.90

Table 3.4 Protein-free chemically defined medium NCTC 109. (Continued)

Vitamins			
Thiamine hydrochloride (Vit B1)	0.025	Choline Chloride	1.25
Riboflavin	0.025	i-Inositol	0.125
Pyridoxine hydrochloride (Vit B6)	0.0625	p-Aminobenzoic Acid	0.125
Pyridoxal hydrochloride	0.0625	Vitamin B12	1.00
Niacin	0.0625	Vitamin A (Crystalline alcohol)	0.25
Niacinamide (Nicotinamide)	0.0625	Calciferol (Vitamin D)	0.25
Pantothenate, calcium salt dextrorotatory	0.025	Menadione (Vitamin K)	0.025
Biotin	0.025	Alpha tocopherol phosphate, disodium salt (Vitamin E)	0.025
Folic Acid	0.025		

Table 3.4 Continued

It should be self evident that these two completely synthetic media are very complex and difficult to prepare. They also probably contain many chemicals which are not required by cells and there have been many attempts to produce simpler formulations. Perhaps the most successful of these are based on Eagle's medium (1955), (see Table 3.5).

minimum
essential
medium,
Eagle's medium

Eagle's medium does not, however, support cell growth in the absence of serum. Out of this type of formulation grew the concept of the 'minimum essential medium' (MEM). Thus we can describe Eagle's medium as a 'Minimum Essential Medium Eagle'. You must, however, realise that the minimum essential ingredients are not identical for all cell lines and manufacturers usually supply a variety of media based on MEM. Some are modified so that they can be autoclaved, some are specially designed for suspension cultures.

Compound	Concentration (mg l^{-1})
L-Arginine	105
L-Cystine	24
L-Histidine	31
L-Isoleucine	52
L-Leucine	52
L-Lysine	58
L-Methionine	15
L-Phenylalanine	32
L-Threonine	48
L-Tryptophan	10
L-Tyrosine	36
L-Valine	46
L-Glutamine	292
Choline	1
Nicotinic acid	1
Pantothenic acid	1
Pyridioxal	1
Riboflavin	0.1
Thiamine	1
L-Inositol	2
Folic acid	1
Glucose	2000
NaCl	8000
KCl	400
CaCl$_2$	140
MgSO$_4$.7H$_2$O	100
MgSO$_4$.6H$_2$O	100
Na$_2$HPO$_4$.2H$_2$O	60
KH$_2$PO$_4$	60
NaHCO$_3$	350
Phenol Red	20
Penicillin	0.5

Table 3.5 Eagle's medium, based on Hanks BSS (can also be used in combination with Earle's BSS for use with 5% CO_2 in a gas phase).

From its original formulation, Eagle's medium has, therefore, been greatly modified. The most common adaption has been to increase the concentration and/or variety of amino acids in the mixture. This generally leads to an improvement in the yield of cells grown in this medium.

Although 'Minimum Essential Medium' based on the formulation of Eagle provides the most commonly used media, other formulations also have important specific uses. We illustrate this by describing two examples.

Fischer's
medium for
leukaemic cells
RPMI medium
for leucocyte-
derived cells

Fischer's medium contains high concentrations of folic acid and is particularly well suited for the cultivation of leukaemic cells. RPMI media were specifically developed to cultivate leucocyte-derived cell lines. Of particular importance is the so called HAT medium. HAT medium is a modification of RPMI medium which enables us to select hybrid cells produced by the fusion of B-cells with myeloma cells (so called hybridomas).

the selective
action of HAT
medium

We will be discussing hybridomas in greater detail in a later chapter (Chapter 7) so we will not discuss this further, we will only outline how HAT works. HAT medium is designed so that it does not support the prolonged growth of B-cells nor does it support the growth of particular myeloma cells which are defective in some enzyme(s) involved in nucleotide biosynthesis. HAT is a mixture of hypoxanthine, aminopterin and thymidine. Aminopterin is a potent inhibitor of nucleotide biosynthesis. This pathway can, however, be by-passed if cells are provided with hypoxanthine and thymidine. Thus spleen cells can grow in HAT medium, but the myeloma cells used in producing hybridomas having a specific metabolic defect cannot use hypoxanthine and thymidine to by-pass the aminopterin inhibition. Thus B-cells will survive but the myeloma cells die in HAT medium. The B-cells die in culture naturally within about 1-2 weeks. If, however, we fuse a B-cell with a myeloma cell, then the myeloma cell acquires the genes that code for the enzymes which enable these cells to by-pass the aminopterin inhibition. Thus fused B-cell:myeloma cells will survive and grow in HAT medium. We have represented this situation in Figure 3.1.

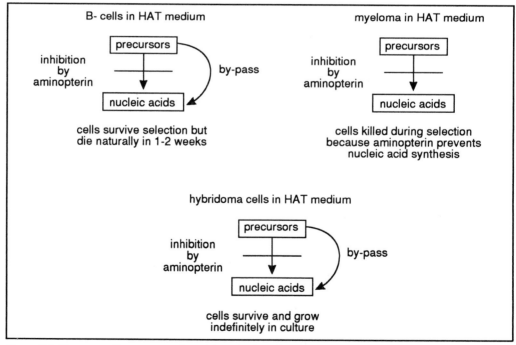

Figure 3.1 Diagrammatic representation of the selection of hybridoma cells in HAT medium.

HAT medium is an excellent example of a medium designed to achieve a particular objective. Below we provide a brief discussion of media used for the culture of cells depending on their origins.

3.4.3 Primary cell culture medium

Most of the media described so far were tested and developed using established cell lines. Media for culturing primary cell lines need a complete amino acid mixture, extensive vitamin supplements, co-enzymes and metabolic intermediates. All the media for primary cell culture require serum to be added for adequate growth.

3.4.4 Media for culturing cold blood vertebrate lines

balanced salt solutions for cold blood vertebrates

Cold blood vertebrate cell lines generally need fewer media supplements and grow in quite simple BSS mixtures with low levels of serum or embryo extracts. Embryonic cell lines require few extraneous nutrients because they have their own food stores. Examples of balanced salt solution used for cold blooded vertebrates are given in Table 3.6.

Substances	Amphibia			Fish	
	Holtfreter	Ringer	Cortland	Holmes & Scott	Keynes & Martins Ferreira
	(g l^{-1})	(g l^{-1})	(g l^{-1})	(g l^{-1})	(g l^{-1})
NaCl	3.50	6.5	7.25	7.41	9.88
KCl	0.05	0.14	0.38	0.37	0.37
CaCl$_2$.2H$_2$O	0.10	0.16	0.23		0.44
MgCl$_2$.6H$_2$O					0.30
MgSO$_4$.7H$_2$O		0.39	0.23	0.31	
NaH$_2$PO$_4$.H$_2$O			0.41	0.40	0.04
Na$_2$HPO$_4$.2H$_2$O				0.20	0.21
KH$_2$PO$_4$				0.17	
NaHCO$_3$	0.20	0.20	1.00	0.31	
Glucose			1.00		

Table 3.6 Balanced salt solutions for cold-blooded vertebrates.

We have included these examples for illustrative purposes only. Many other formulations have been used and it is wise to consult the literature and manufacturers specifications before beginning work with these types of cell lines.

3.4.5 Media for culturing invertebrate tissues

The defined components of the media used for culturing invertebrate tissues are similar to those used for mammalian cells. Medium 199 (Table 3.2) is used extensively to culture insect cell lines. The pH should be kept slightly acidic, usually between 6.3-6.9. The osmotic pressure requirements may also vary a little. Insects feeding on animals require slightly higher osmotic pressures than cell lines from insects which feed on plants. Insect blood sometimes has very high amino acid levels which can significantly increase osmotic pressures.

Wyatt's medium, for example, takes this into account. By basing its composition on the analysis of the haemolymph of silkworms, Wyatt was able to produce medium giving good growth conditions for silkworm cell lines. In 1958 Grace improved Wyatt's medium by adding more B vitamins. Grace's insect medium is now commonly produced by commercial suppliers along with those of Schneider and Rinaldini. These insect cell culture media are commercially available and details of their formulation and application are available through commercial suppliers.

Many synthetic or defined media are prepared as 10 x concentrated solutions which can be stored in the cold room or refrigerator without occupying a lot of space. The concentrated medium is usually diluted before use in sterile, double distilled water or in a BSS. We remind you that the choice of BSS used to make up defined media depends on:

- the type of atmosphere in which the cells are to be grown. (For example Earle's BSS is used as a diluent for Eagle's medium which is used for equilibration with CO_2. Enriched Hank's BSS is used when equilibration with air is required);

- whether the medium is to be used for tissue disaggregation or monolayer dispersal. It may be necessary to use medium from which calcium and magnesium ions have been omitted for these purposes. Moscana's Ca^{2+} and Mg^{2+} free saline (CMF) or Dulbecco's phosphate buffered solution A (PBSA) are suitable for these purposes;

- whether the medium is to be used for growing cells which adhere to a substrate or for growing cells as suspension cultures. Reducing the calcium ions in the medium helps to reduce cell aggregation and attachment. Minimum Essential Medium (MEM) (a variant of Eagle's medium) has a reduced calcium ion concentration. Specific adhesions factors may be added to the medium. We will discuss these in more detail later.

SAQ 3.4

1) Look back at Table 3.1, what do you think would be the result if the concentration of calcium chloride in Simms medium was increased five-fold and the media was sterilised by autoclaving?

2) In Joklik's modification of Minimum Essential Medium (Eagle) the medium does not contain any calcium chloride. Suggest the main use of Joklik's medium.

3) You have become interested in studying kidney cells derived from cats. This study will involve the cultivation of the kidney cells. A commercial supplier of tissue culture media provides you with a list the following media:

a) Dulbecco's modified Eagle's medium.

b) NCTC 109 medium.

c) Minimum Essential Medium Eagle.

d) RPMI medium.

e) MCDN 105 medium (originally developed for human fibroblast-like cells).

f) MCDB 153 medium (originally developed for cloning and long term growth of human keratinocytes).

g) MCDN 101 medium (originally for chicken embryo fibroblasts).

h) MCBB 302 medium (originally for growth of Chinese hamster ovary cells).

i) NCTC 135 medium (originally for mouse L-cells).

j) SFRE 199 medium (originally developed for baboon kidney cells).

Sera, from:

Foetal bovine, bovine
Foetal horse, horse
Chicken
Goat
Human
Porcine
Rabbit
Sheep

From this list, make a selection of media you might try to use to cultivate these kidney cells.

3.5 Further considerations in media formulation

Now that we have introduced you to a variety of media, we can focus on some of the factors which must be considered in formulating and preparing media. In this section we will examine:

- the solubility of material;

- purity of materials;

- chemical instability;

- water and osmolarity;

- temperature;

- viscosity and surface tension.

In order to get the best out of your media, you need to be alert to possible reactions between ingredients and to the factors which influence the stability of chemicals. We cannot go through all of the possibilities but, instead, we alert you to the major issues.

3.5.1 Solubility of materials

problems with tyrosine, lipids Ca^{2+}

Many of the components of synthetic media are very sparingly soluble in water. For example, tyrosine is only soluble in acid solutions. Lipids have to be dissolved in alcohol-based solutions first before they can be mixed with water. Calcium and phosphates cannot be mixed except in very dilute solutions otherwise the calcium salts will be precipitated out of solutions.

3.5.2 Purity of materials

toxicity of metal ions

Tissue culture grade reagents must be used when formulating media. Even very low concentrations of heavy metal ions are toxic to cells, and must be absent from salts used for making media. Other minerals such as zinc, manganese and cobalt are essential for cell growth but are required in very small amounts. Often they can be found in adequate concentrations as contaminants of other salts.

3.5.3 Chemical instability of materials

Glutamine keeps well as a dry powder, but once made up in solution it must be kept frozen to prevent degradation. It is the most unstable of the essential amino acids and decomposes rapidly to form pyrrolidine carboxylic acid and ammonia in solution. It must be added last to the complete medium and replenished regularly. Many components of the medium cannot be heated together, for example, if cystine and ascorbic acid are autoclaved in the same solution, the ascorbic acid becomes oxidised and destroyed. Glucose and ascorbic acid are both destroyed in alkali solutions.

3.5.4 Water

need for pure water

Ordinary tap water contains a lot of ions which are toxic to cells. It may also contain endotoxins released by the growth of Gram negative bacteria within water mains. For media formulation, it is essential to use glass distilled or double de-ionised water. (De-ionised water is filtered through a mixed bed resin and then distilled through glass). Glass distilled water should be stored in plastic containers if possible in order to prevent leaching of metallic ions from the glass.

3.5.5 Osmolarity

Robust, established cell lines can tolerate wide fluctuations in osmolarity. Cells in primary cultures are, however, very sensitive and high yields of cells can only be obtained if the osmolarity is carefully controlled. Cells usually need osmotic pressures of between 280-320 mOsm/litre and, ideally, the osmotic pressure should be maintained at around 290-300 mOsm/litre.

If HEPES, acid or alkali are used to adjust the pH of the medium then an equivalent amount of sodium chloride should be omitted to allow for the increase in osmotic pressure. The correct amount of NaCl to be omitted can calculated using formulae based on:

$$\frac{D-O}{32} = X$$

Assuming that $1 \text{ mg NaCl ml}^{-1} = 32$ mOsm increase/decrease

D = desired (mOsm) osmotic pressure

O = observed (mOsm) osmotic pressure

X = ml of stock NaCl to be added or omitted to obtain optimum osmotic pressure (providing that 1 ml of stock changes the concentration of NaCl in the medium by 1 mg ml^{-1}).

Let us try a calculation

$\prod$ Assume that you have a stock solution of sodium chloride. If you add 1 ml of this solution to a medium, it increases the sodium chloride concentration of the medium by 3 mg ml^{-1}.

If the medium has an osmotic pressure of 200 mOsm l^{-1}, how much of the stock sodium chloride solution needs to be added to increase the osmotic pressure to 300 mOsm l^{-1}. (Try this on a piece of paper before reading on).

The answer is approximately 1 ml.

Since an increase of 1 mg NaCl ml^{-1} leads to 32 mOsm increase in osmotic pressure, then 3 mg NaCl ml^{-1} leads to approximately 96 mOsm increase in osmotic pressure.

Thus the relationship described above can be modified to

$$\frac{D-O}{96} = X$$

where;

D = desired osmotic pressure

O = observed osmotic pressure

X = ml of stock NaCl which increases the concentration of NaCl by 3 mg ml^{-1} for every ml added.

Substituting in D = 300, O = 200, then $X = \dfrac{100}{96}$ = approximately 1 ml.

Osmolarity is usually measured by freezing point depression or elevation of vapour pressure. Commercially produced media are carefully adjusted to provide optimal buffering and osmolarity so this formula needs to be used only when making up media from scratch.

3.5.6 Temperature

temperature influences pH

The direct effects of temperature on cells are obvious, but the temperature also effects the pH of the medium due to increased solubility of CO_2 at lower temperatures. The pKa of the buffer and the degree of ionisation of serum components is also affected by the temperature. The pH of a typical medium at room temperature is around 0.2 units lower than at 37°C. Phenol red is added to media as a pH indicator. It can, however, be slightly toxic even at 0.005% concentration if serum is not present in media.

3.5.7 Viscosity

viscosity and serum content

The viscosity of the medium is mainly influenced by its serum content, and usually has little effect on cell growth. It is important, however, when the cell suspension is agitated (for example, when a culture is stirred or when cells are disrupted after trypsinisation). Damage to cells in large scale cultures can be reduced by adding carboxymethyl cellulose or polyvinyl pyrollodine to increase viscosity. They are especially important when low levels of serum are used for culturing cells in scaled up stirred reactors (see Chapter 8).

3.5.8 Surface tension and foaming

Lowering the surface tension in the medium can be used to promote adherence of cells to the substratum when this is required.

problems of foaming

Foaming occurs especially in medium containing serum and can cause the denaturation of proteins and inactivation of enzymes. If the foam reaches the neck of the culture vessel, microbial contamination can also become a problem.

Silicone antifoaming agents are used in suspension cultures where CO_2/air mixtures are bubbled through the medium.

SAQ 3.5	Carboxymethyl cellulose is to be used to increase the viscosity of a medium. Would you add this to the medium before, or after, sterilisation? (Give your reasons).

3.6 Nutritional components of media

In this section we will briefly consider the nutritional components of media. For this we will divide the discussion into:

- amino acids;

- nucleic acid precursors;

- carbon sources;

- vitamins.

In the final part of this section we will briefly describe the role of a variety of polymers used in some media.

3.6.1 Amino acids

essential and other amino acids

Essential amino acids are those which cannot be synthesised from raw materials by heterotrophic organisms. These amino acids, together with cysteine and tyrosine must be incorporated into a chemically defined synthetic medium even if a serum supplement is to be used to grow the cells. Individual requirements may vary and non-essential amino acids may need to be added to compensate for particular cells' inability to synthesise adequate amounts of them or because they become leached out into the medium. The concentration of amino acids influences cell yield and also affects survival and growth rates. Amino acid deficiency inhibits cell division, induces chromosomal damage and increases lysosomal activity and cell death. Imbalances of amino acid concentrations may also produce karyotype changes.

The most rapid use of amino acids occurs during the lag phase of the growth cycle, and cystine, glutamine, isoleucine and serine are used up very quickly. Glutamine is used by most cells (though some prefer glutamic acid) and provides an essential source of carbon and energy. It is the most unstable (*labile*) of the amino acids and needs to be replenished by regular medium changes. It is also likely that glutamine has a role in the formation of molecules involved in cell adhesion.

3.6.2 Nucleic acid precursors

Adenosine, guanosine, cytidine, uridine and thymidine are often added to media, particularly if folic acid is in short supply such as in low density cultures. Medium 199 is usually also deficient in folic acid and needs nucleic acid precursors supplements.

3.6.3 Carbon sources

The growth of cells depends on the availability of carbon sources, usually glucose and glutamine. If glucose is used, a pyruvate supplement is also necessary.

3.6.4 Vitamins

role as cofactors

You will be aware that many vitamins are required to act as cofactors in metabolism (eg Niacin $\rightarrow$ NAD(P)$^+$; riboflavin $\rightarrow$ FMN, FAD). We will not re-examine these again here. However, we will describe some other important functions of vitamins in cell culture.

retinoids
choline
ascorbate

Retinoids help cells to adhere to substrata and can be used at relatively high concentrations (1 mg ml^{-1}) for initial cell adhesion; and then at lower levels to maintain the culture. A sufficient supply of choline is essential for incorporation into membrane phospholipids. Deficiency of this vitamin causes rounding up of cells and eventual death. Choline is normally provided in serum and needs only to be supplemented if serum-free medium is used. Ascorbic acid (vitamin C) is necessary for most cells but should be omitted from media used for growing RNA tumour viruses.

3.6.5 Polymers

protection
against
mechanical
damage

These are high molecular weight components of medium and are necessary for growing cells at low densities or cells which have low plating efficiencies. Polymers improve cell survival by protecting against mechanical damage, but do not themselves contribute to growth. Ficoll 400, Dextran T70, Dextran T-500 and methylcellulose have all been used as additions to media.

3.7 The role of serum in cell culture

Serum is a very complex mixture of many small and large molecules with growth promoting and growth inhibiting activities. Table 3.7 describes the roles of some serum components.

Serum is used in cell culture medium to provide:

- hormonal factors which promote cell growth and cellular functions;

- attachment and spreading factors;

- transport proteins carrying hormones, minerals, lipids etc.

These are summarised in Table 3.7 and discussed in a little bit more detail below.

3.7.1 Growth factors

patterns of
growth factor
requirements

Most growth factors are available in minute amounts in serum, usually in the order of nanograms or picograms per ml. Some, like the colony stimulating factors, act specifically on cells at a distinct stage of differentiation. Others act synergistically on several cells types, for example epidermal growth factor (EGF) promote the proliferation of fibroblasts, epidermal and glial cells. Similarly, a single cell type may be influenced by a variety of growth factors for example fibroblasts respond to fibroblast growth factor (FGF), EGF, platelet-derived growth factor (PDGF) and somatomedins. The growth factor requirements of cells are extremely complex and our lack of knowledge often forces us to use complex mixtures such as those supplied by serum rather than using defined mixtures of purified chemicals.

Component	Function
Proteins	
Albumin	carries lipids, hormones and minerals and provides osmotic pressure
Fibronectin	promotes cell attachment to substratum
Fetuin	enhances cell attachment
Transferrin	binds iron
α macroglobulin	inhibits trypsin
Hormones and growth factors	
Insulin	uptake of glucose and amino acids
Platelet-derived growth factor (PDGF)	mitogen for fibroblasts, smooth muscle cells etc
Fibroblast growth factor (FGF)	mitogen - growth factor
Endothelial growth factor (ECGF)	mitogen - growth factor
Epidermal growth factor (EGF)	mitogen - growth factor
Hydrocortisone	promotes cell attachment
Steroid hormones	mitogens - growth factor
Thyroid hormones (T_3, T_4)	oxygen consumption, metabolic rate control, growth and differentiation of various cells
Lipids	
Cholesterol	membrane synthesis
Linoleic acid	
Prostaglandins	
Metabolites	
Amino acids	cell proliferation
Polyamines	

Table 3.7 Some components of serum and their role in cell survival and growth.

3.7.2 Hormones

Insulin is essential for the growth of nearly all cells in culture. It is very sensitive to inactivation by cysteine and has a very short half life so large amounts need to be added to the culture medium. Glucocorticoid hormones such as hydrocortisone and dexamethesome can either stimulate or inhibit cell growth depending on cell type. Other steroid hormones (such as oestradiol, testosterone and progesterone) and thyroid hormones are also necessary to support some cell lines. Again we are often faced with difficulties in predicting the exact hormonal requirements of cells.

hormone requirements may be difficult to predict

3.7.3 Attachment and spreading factors

adhesion factors, chondronectin, laminon fibronectin

Most non-transformed cells have to attach themselves to a solid substrate in order to grow. Only haemopoetic and transformed cells can multiply without attachment. Chondronectin is required for the adhesion of chondrocytes and laminin for the adhesion of epithelial cells. Fibronectin is also involved in cell attachment.

3.7.4 Binding proteins

albumin and
transferrin as
transport
proteins

Factors which are of low molecular weight are often carried into cells by binding onto transport proteins. Examples of transport proteins are, albumin which carries vitamins, lipids and hormones into cells and transferrin which is involved in the binding and transport of iron.

3.7.5 Fatty acids

Cells require varying amounts of essential fatty acids, phospholipids and cholesterol. Prostaglandins E and $F_{2\alpha}$ are also involved in cell growth, possibly acting in conjunction with EGF and other growth factors.

3.7.6 Trace elements

Copper, zinc, cobalt, manganese, molybdenum and selenium are present in serum and are thought to be involved in activating enzymes and protecting against free radicals which cause damage to DNA.

∏ Serum, therefore, provides an enormous variety of the necessary ingredients to successfully cultivate animal cells. It is not surprising that it finds extensive use as a supplement in a wide variety of media. However, its use does present many potential disadvantages. Before reading on see if you can make a list of these disadvantages.

The sort of list we hope you have generated should include the following items:

* sera from different animals differ widely in their properties. Even batches of sera from the same species vary enormously, so all serum supplements have to be batch tested extensively from time to time. This procedure is both time consuming and expensive;

* under normal physiological conditions, most cells do not come into contact with serum except during wound healing, so it is not the most natural medium for culturing cells and this can complicate the process of extrapolating *in vitro* experimental results to the *in vivo* situation;

* some components of serum are actually cytotoxic under some conditions. Serum often contains selective inhibitors, bacterial endotoxins and lipids, and polyamine oxidase which reacts with polyamines (spermine, spermidine) produced by proliferating cells to form toxic polyaminoaldehydes which can suppress cell yields;

* specific growth factors may be present in inadequate amounts to grow some cells and purified growth factors may still have to be added;

* foetal calf serum, which is used for many of the more demanding cell types is very expensive and limited in supply. It needs to be used at concentrations of between 5-20% depending on the cell type and this can make it prohibitively expensive. Hormone levels may vary and foetal calf serum often contains very high levels of the enzyme arginase which depletes the culture of the essential amino acid arginine. Sera from newborn and adult animals are used but often the latter have high levels of gammaglobulins (antibodies) which may cause subsequent problems.

serum
substitute
We are, however, in quite a fortunate position. The development of biotechnological enterprises means that a wide variety of cell culture medium supplements are now available and this alleviates the need to use serum so extensively.

Π Make a list of the advantages that are associated with using low serum or serum-free media.

The main advantages are:

• reproducibility between cultures is improved;

• the risk of introducing contaminants (bacteria, viruses, mycoplasma) is reduced;

• serum cytotoxicity and protein interference is reduced;

• serum - free media can be less expensive;

• culture products are easier to purify;

• fibroblast overgrowth in primary cultures is prevented (serum promotes fibroblast growth in wound healing).

Low serum and serum-free media are particularly valuable where it is vital to control the culture conditions precisely. If, for example, the upregulation of immunological markers on cells is being studied, it is important to eliminate the undefined growth factors present in serum. These reasons also apply to the large scale cell culture processes for growing hybridomas to produce monoclonal antibodies or lymphomas to produce lymphokines and other factors.

serum-free
medium often
very specific
A problem with most serum-free media is that they are often highly specific to one or two cells types, so a different formulation for different types of cell lines is required. We will consider this in a little more detail in the final section of this chapter.

It should, however, be borne in mind that not all of the components of serum are known, nor their functions completely understood, and so artificial serum substitutes are still inadequate for very long term culture.

3.8 Choosing a medium for different cell types

Chemically defined serum-free cell culture medium should be bought or designed with the type of cells to be cultured in mind. A number of questions need to be asked:

• are the cells normal (non-transformed) or transformed?

• is the culture intended for survival or growth?

• is the culture comprised of a mixed populations of cells (eg fibroblasts and epithelial cells)?

• is differentiation or proliferation of the cells required?

• are the cells to grow anchored or in suspension?

Suppliers of tissue culture media offer a range of growth factors, attachment factors, matrix factors, hormones and biological buffers. They also offer various serum replacements. The most commonly available growth factors are for endothial cells, epidermal cells, fibroblasts, lymphocytes (especially T cells) and nerve cells.

As our knowledge of the processes which regulate cell growth and division in animal cells increases and we understand more about the control of cellular differentiation, it will become possible to make even better formulations to achieve particular objectives.

| SAQ 3.6 | Below are two lists. In one list we have included a variety of cell types and in the other, growth factors. For each cell type select a growth factor you would anticipate would be needed in the culture medium. (If your knowledge of cell biology and animal physiology is limited you might find this question difficult. In this case read our response). |

Cell type/source	Growth factor
a) T cells	FGF (fibroblast growth factor)
b) skin explant	NGF-β (nerve growth factor β)
c) liver cells	interleukin-2
d) fibroblasts	epidermal growth factor
e) ganglia	gly-his-lys

Summary and objectives

In this chapter we have described the essential features of the media used to cultivate animal cells *in vitro*. We explained that media could be composed of natural materials such as plasma or be formulated using purified chemicals. Within the chapter we explained that media need to supply more than just the nutrients required for growth. With most animal cell lines there are also requirements for growth factors, hormones and (for anchorage-dependent cells) attachment factors. Traditionally these were supplied using serum to supplement media but increasingly we are able to use totally synthetic (defined) media.

We also explained the advantages of using defined media over serum-based media.

Now that you have completed this chapter you should be able to:

- explain the role of the various constituents of media and supplements which are used in tissue culture;

- describe the need for certain physical properties of media to be maintained eg pH, osmolarity, temperature, viscosity, surface tension and foaming;

- list a large number of the essential chemical constituents of media including, salts, amino acids, vitamins, minerals, organic supplements, hormones, growth factors and serum;

- explain the role of serum as a media supplement and discuss the advantages and disadvantages of using serum as a supplement;

- discuss the importance of using complete, synthetic media without serum supplements;

- describe the factors influencing the choice of media for different types of cell culture;

- select appropriate media for the cultivation of a particular cell type.

Characterisation of cell lines

4.1 Introduction   74

4.2 Species verification   74

4.3 Intra-species contamination   78

4.4 Characterisation of cell type and stage of differentiation   81

4.5 Microbial contaminations   93

Summary and objectives   98

Characterisation of cell lines

4.1 Introduction

Once a cell line has been established, it has to be characterised. The most important reasons for this are to confirm the species and tissue of origin, and to determine the differentiated status of the cell within the lineage. There is also a need to check that there has been no cross-contamination which is particularly a problem since fibroblasts tend to grow well in culture and may contaminate other cell types. Even once a cell line has been extensively characterised, there is a very real possibility of contamination with other cell lines. Much time and money has been wasted as a result of working with cultured cells which were derived from the wrong species or the wrong tissue because, at some stage, a well-characterised cell line had become contaminated with another cell line. In fact, data from a large number of studies suggest that as many as 35% of cell lines used are contaminated by another species or another cell line of the same species. Cell lines should, therefore, be periodically checked against this and monitored for instability and variation. We will consider all of these points in this chapter. Finally we will consider the types of experiments which are performed to check that the culture has not become contaminated with micro-organisms.

4.2 Species verification

If you were working with a specific cell line, you would usually know the species of origin of that cell line, since you would know from which organism the cells had been isolated. However, there is a need to verify that the cells are indeed derived from that particular species, since there is always a risk of cross-contamination. The species of origin can be confirmed by a number of methods (cytogenetics, isoenzymology and by a variety of immunological tests) which we will discuss below.

the need to
confirm
species origin
of cell lines

4.2.1 Cytogenetics

The most characteristic trait of a species is its chromosome content and chromosomal analysis provides the most reliable method of distinguishing between species. The correlation of chromosomal structure with heredity and variation is referred to as cytogenetics and the word karyotype is a collective term used to describe the chromosome number, size and shape. Karyotypic techniques are those techniques which rely on the analysis of karyotype.

karyotype

Karyotype techniques

Since chromosome shape and size varies with the stage of nuclear division, it is important to analyse the karyotype at a specific stage of the cell cycle. The stage which is usually studied is the metaphase stage of mitosis. Cells are arrested in metaphase using agents such as colcemid or vinblastine, and, in the case of anchorage-dependent cells, brought into suspension with trypsin. They are then swollen in hypotonic medium and fixed using an ice-cold mixture of acetic acid and methanol. The cells are

metaphase
chromosomes
used

then dropped onto a slide, stained with a dye called Giemsa dye, and examined under the microscope (see Figure 4.1).

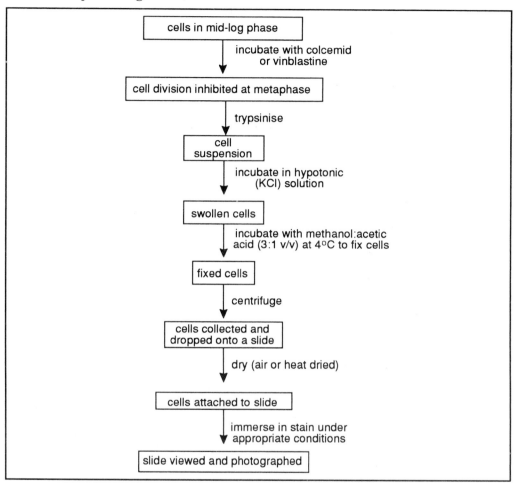

Figure 4.1 Stages in staining chromosomes for karyotyping.

A group of techniques has been devised to enable individual chromosomes pairs to be identified. The use of Giemsa dye outlined above is only one such technique and is sometimes called G-banding. Other techniques include the use of quinacrine mustard, which stains the interband regions, and C-banding which stains the centromere region. These other techniques are frequently used to clarify an area of uncertainty arising from G-banding. Quinacrine banding (Q-banding) employs the use of quinacrine mustard or quinacrine dihydrochloride. Staining with these reagents produces bright and dull fluorescent bands closely resembling the patterns given by G-banding. This technique is, however, of particular value in analysing Y chromosomes in humans. With Q-banding, the Y chromosomes are strongly fluorescent and are very distinctive whilst G-banding produces rather dull uniformly stained Y chromosomes.

G-banding
C-banding

Q-banding

G11 banding

Another refinement of chromosome staining is to use Giemsa staining at pH11 instead of the usual pH6.8. This technique, called G11 banding stains mouse chromosomes dark magenta with pale blue centromeres. Under the same conditions, human chromosomes stain blue with red regions on a number of chromosomes (chromosomes 1, 4, 5, 7, 9, 10, 13, 14, 15, 20, 21 and 22). This technique is, therefore, of value in examining human/mouse hybrids (for example hybridomas).

You should realise that the actual protocols used to stain cells are slightly modified for different cell lines in order to produce satisfactory staining. For example, the hypotonic medium used to swell human fibroblasts is 0.07 mol l^{-1} KCl for 10 minutes whilst Chinese Hamster cells require only a brief exposure (3 minutes) to 0.1 mol l^{-1} KCl to achieve satisfactory swelling.

∏ To produce cells that are in metaphase, how long should cells be exposed to colcemid or vinblastine?

The answer is that it depends upon the doubling time of the cells. Fast growing cells will all accumulate at metaphase quite quickly, whereas at slower growth rates, a longer incubation would be required to achieve the same results.

∏ Care has to be taken not to expose cells for too long to colcemid of vinblastine. What do you think would be the consequences of a long incubation in the presence of these mitotic inhibitors?

If cells are incubated in these presence of these reagents, the cell cycle is halted at metaphase. The chromosomes continue to condense and eventually become so short that it is difficult to distinguish the banding patterns in the chromosome arms.

It is important to note that after staining, 'aged' slides (3-7 days) often give clearer patterns than freshly stained slides.

analysis of karyotypes

To construct the karyotype, photographs must be taken of the image seen through the microscope. The chromosomes in these 'photomicrographs' are cut out and arranged according to arm length, position of the centromere and presence of secondary constrictions. This can then be compared to textbooks showing conventionally stained preparations of different species.

4.2.2 Isoenzymology

isoenzymes (isozymes)

Some enzymes exist in animal tissues in multiple forms, and different molecular forms of an enzyme catalysing the same reaction are called isoenzymes or isozymes. Isoenzymes can be separated chromatographically or electrophoretically and the number and intensity of enzyme bands are indicative of the species and lineage of the cell line. In a well established method, only 1 µl of clarified cell extract is required for each enzyme to be analysed. The cell extract is placed into pre-cut slots in an agarose gel which fits into a constant voltage cassette system. After electrophoresis, the gel is stained and visualised by one of a variety of protocols, depending on the enzymes being analysed. The bands obtained can be compared to publications showing conventional patterns of banding for different species.

Determination of the mobilities of different isoenzyme systems by the agarose cassette system is a very effective way to establish the identity of a cell line. It is usually accurate, and it is quicker and less expensive than chromosomal analysis. Three particular isoenzymes are routinely examined in order to identify the species of origin with a high degree of certainty: glucose 6-phosphate dehydrogenase (G6PDH), lactose dehydrogenase (LDH), nucleoside phosphorylase (NP). The identification into different species can be accomplished largely by interpretation of LDH and G6PDH isoenzyme mobilities. However, these two enzymes do not distinguish amongst several primate species. The enzyme NP has, however, proved useful in differentiating between some primate cell lines. No isoenzyme has yet been successful in separating chimpanzee from human cell lines.

G6PDH, LDH and NP often used

Let us consider a specific example of how isoenzymes may be helpful in detecting contamination by undesirable cell lines.

G6PDH type A and B in humans

Amongst the human population there are two isoenzymes (A and B) with glucose-6-phosphate dehydrogenase activity. Type A is found in Caucasians and 80% of Negroids. 20% of Negroids, however, produce isoenzyme B. HeLa cell lines produce typeB glucose-6-phosphate dehydrogenase and thus can be used to detect the contamination of many human cell lines by HeLa cells. Alternatively the presence of type A G6PDH in HeLa cell lines would indicate that these had become contaminated by other human cell lines.

∏ Does the absence of typeA G6PDH in a HeLa cell culture prove it is not contaminated by other cell lines?

The answer is no. If such a culture was contaminated by cells derived from a Negroid which produced type B G6PDH, then this contamination would not be detected by measuring the G6PDH isoenzyme pattern of the HeLa cell culture.

4.2.3 Immunological tests

fluorescent antibodies

Species-specific antigens can be detected by immunofluorescence. The antisera are coupled with fluorescein isothiocyanate and mixed with the test cells, then the cells are incubated, washed and mounted on slides. The cells are examined using a UV fluorescent microscope - a positive reaction is denoted by a bright green peripheral fluorescence of the cell membrane. Although this is the simplest diagnostic test, it is not always practically feasible.

Similar techniques may be also used to detect the presence of mycoplasma or viruses in the culture. We will examine these techniques a little later.

The key to success with these techniques depends upon

- the specificity of the antisera;

- the intensity of the stain.

With the development of monoclonal antibody production by 'immortalised' cell lines, such diagnostic procedures are becoming more reliable and of wider application.

4.3 Intra-species contamination

Once the species of origin of a cell line has been confirmed, it may be important to find out which individual within that species the cell line is derived from. Again a number of techniques are available. These techniques are dependent upon the fact that under most circumstances the genetic information of any individual is unique. Below we discuss two methods of analysis which involve examination of the genetic material directly and three methods which involve investigation of the protein products of the genetic material.

4.3.1 Cytogenetics

The banding patterns mentioned in 4.2.1 are characteristic for different members of any species, so often differences between individuals can be recognised by comparing karyotypes. However, this method is relatively crude compared to the recombinant DNA methods which have now become standard practise in many laboratories.

4.3.2 Recombinant DNA methods

restriction fragment length polymorphisms

It is possible to directly determine an individual's genotype by typing the genes themselves. For many genes there are slight variations between individuals of a species, and such molecular differences in the DNA of a particular gene are called polymorphisms. The presence of dissimilar arrangements of DNA bases means that each individual's DNA is cut at slightly different sites by restriction enzymes, generating distinct lengths of restriction fragments. DNA probes can be used to detect polymorphisms in a technique known as restriction fragment length polymorphism (RFLP). You may have heard of this technique from television, radio or newspaper reports since RFLP mapping has been used for diagnosing inherited diseases, and also in forensic science, for example to confirm or refute the presence of an individual at the scene of a crime. It has also been used in paternity testing - since the technique can give conclusive evidence in cases where the identity of a father is questioned.

DNA fingerprinting

Basically the technique is as follows. Cellular DNA is digested with restriction enzymes and the fragments of DNA are separated by agrose gel electrophoresis. The DNA is then transferred to a membrane by Southern blotting and the DNA on the membrane is 'melted' to form single strands. The membrane is then exposed to radioactively labelled DNA probes. These probes are prepared using highly repetitive nucleotide sequences (so called satellite DNA). These radioactive probes will react with the DNA fragments which contain complementary nucleotide sequences. The presence of radioactive probes is detected by autoradiography. This process, often called DNA fingerprinting, is illustrated in Figure 4.2.

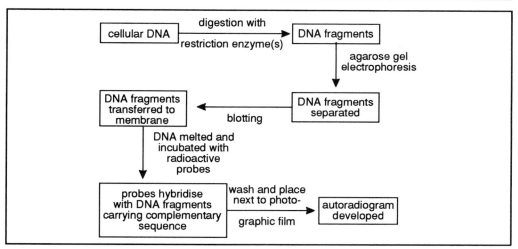

Figure 4.2 The principles of DNA fingerprinting (see text for further details).

You should note that this technique is useful for all species and can be used to confirm the identity of hybrid cells.

Using the technique of DNA fingerprinting described in the text, the following autoradiograms were produced.

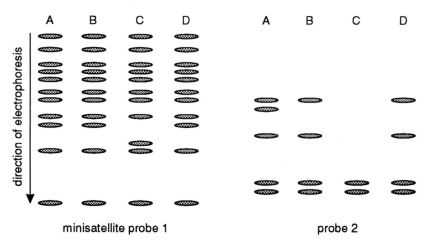

Two different probes were used to detect particular nucleotide sequences.

1) From these results, are the cell lines A, B, C and D identical?

2) If you had to routinely check the identity and purity of these cell lines, would you use probe 1 or probe 2 or both?

4.3.3 High resolution 2-D electrophoresis

There are genetic differences between most individuals of any species which account for the uniqueness of that individual. One of the consequences of these distinctive gene patterns is that there are also characteristic patterns of proteins for each individual. Another approach to characterising cell lines is to study the proteins produced by that cell line, either by analysing the overall protein content or by probing for specific proteins which are known to be highly variable.

separation by isoelectric focusing and SDS gel electrophoresis

The different proteins in the cell can be separated on the basis of isoelectric point and molecular weight using isoelectric focusing in the presence of urea in one dimension, and electrophoresis in the presence of the detergent sodium dodecyl sulphate (SDS) in the second dimension. Isoelectric focusing will separate the proteins according to isoelectric point and electrophoresis of proteins in the presence of SDS separates proteins of the basis of their size. In practice such an approach enables us to separate (resolve) many proteins. Perhaps as many as 10,000 proteins in a single sample can be resolved using this technique. This gives a very specific 'fingerprint' for each cell line.

4.3.4 Allozyme analysis

allozymes

Different individuals of the same species may express different alleles for a given enzyme locus. Allelic isoenzymes are referred to as allozymes, and the analysis of allozymes is like a genetic signature for that cell line. The methods involved in analysis of allozyme expression are similar to the isoenzyme analysis referred to in 4.2.2.

4.3.5 Blood group antigens and HL-A

The final method of analysis of cell lines which we will cover involves the determination of the antigens present on the surface of the cell. Blood group antigens are present on normal human epithelium in primary culture and on some continuous epithelial lines. The use of anti-human A, B or AB typing antiserum can be used to type these cell lines.

major histocompatibility complex (MHCs)

The major histocompatibility antigens (or major histocompatibility complex - MHC) are highly variable antigens which are responsible for the immunological individuality of each person. These antigens are present on the plasma membrane of nucleated cells and are responsible for the fact that tissue transplanted from one individual to an unrelated individual will be recognised as foreign and will be rejected by the recipient's immune system. In Man, the histocompatibility antigens are referred to as the Human Lymphocyte Antigens (HL-A) system. The genes coding for the HL-A antigens are on chromosome 6 and occupy 4 loci (designated A, B C and D) along the chromosome. The genes at any locus are not always the same, and the different forms of each gene are called alleles. The typing of these alleles using a two stage complement-dependent cytotoxicity test provides a specific system for the identification of cell lines. This process is illustrated in Figure 4.3.

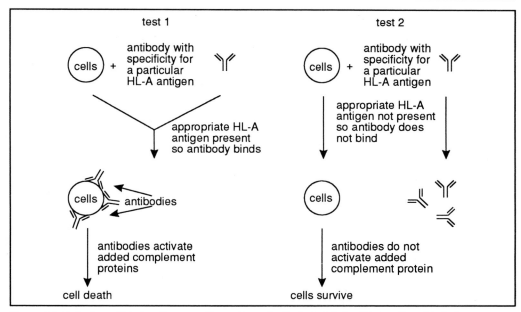

Figure 4.3 Detection of HL-A antigens on the surface of cells. In test 1, the HL-A antigen binds the antibody. The fixed antibody will then interact with complement proteins leading to the death of the cell. In test 2, the HL-A antigens on the surface of the cell do not interact with the antibody so the antibody remains free. The unbound antibodies do not activate complement and so the cells survive. Note that if you are unfamiliar with the roles and activities of antibodies and complement, we recommend the BIOTOL text 'Cellular Interactions and Immunobiology'.

SAQ 4.2	Select an appropriate technique for each of the following.

1) To determine whether or not a culture of HeLa cells are contaminated with other human-derived cells.

2) To check the purity of a Chinese Hamster Ovary (CHO) cell culture after several sub-cultures.

3) To identify human and mouse chromosomes present in a culture of hybrid cells.

4) To confirm if a human cell line has been derived from a particular individual.

5) To confirm that two cell lines probably had a common origin.

4.4 Characterisation of cell type and stage of differentiation

Once the species from which the cell line originates has been determined, the tissue of origin can be confirmed. The technique used will depend on the cell type, but may involve functional assays such as milk production, cytogenetic or isoenzyme analysis, or other methods. Before we start looking at specific tests for determining tissue types, it is important to clarify the major cell types which exist *in vivo* and the processes by which different cell types arise *in vivo*.

4.4.1 The organisation of cells *in vivo*

organisation of cells into tissues and organs

In vivo, cells are organised into tissues (groups of similarly specialised cells and associated extracellular matrix) which perform a specific function or group of functions. Each kind of tissue is composed of cells with a characteristic size, shape and arrangement. Several types of tissue may unite to form an organ and organs are arranged into organ systems. Complex animals have a great variety of organ systems such as the lymphatic system, the respiratory system, the digestive system, the endocrine system. Surrounding the surface of most cells is a complex mixture of glycoproteins and proteoglycans which is highly specific for each tissue.

Animal tissues may be classified as epithelial, connective, muscular, nervous and blood and lymph.

Connective tissue

Connective tissue joins together the tissues of the body, supports the body and protects the underlying organs. Connective tissue consists of relatively few cells embedded in an extracellular matrix. It is also referred to as stroma, and the cells are sometimes called stromal cells. It is important to remember that there are many different types of connective tissue and they include adipose tissue, cartilage and bone. Fibroblasts are the connective tissue cells that produce extracellular matrix components. Examples of connective tissue are shown in Figure 4.4a.

Epithelial tissue

Epithelial tissue consists of cells fitted tightly together to form a continuous layer or sheet of cells covering a body surface or lining a cavity within the body. One surface of the sheet is attached to the underlying tissue by a non-cellular basement membrane composed of tiny fibres of polysaccharide material produced by the epithelial cells. There are different types of epithelium such as squamous, cuboidal and columnar. In addition, epithelial tissues can be classified according to whether they grow as a single layer (simple) or as two or more layers (stratified). Many epithelial cells are specialised to secrete a specific product, such as milk, mucus or sweat. Examples of epithelial tissue are shown in Figure 4.4b.

Muscular tissue

Muscular tissue is composed of cells specialised in contraction. They contain contractile fibres called myofibrils which are composed of actin and myosin. They are further categorised as skeletal, smooth or cardiac according to location and properties of the cells. Examples of muscular tissue are shown in Figure 4.4c.

Nervous tissue

Nervous tissue is composed of cells specialised in conducting impulses and the associated connecting and supporting cells. The brain, spinal cord, peripheral nerves and ganglia all contain nervous tissue. The actual conducting cells of the nervous system are called neurons and they typically consist of a cell body, dendrites (fibres that conduct impulses towards the cell body) and axons (long fibres that transmit impulses away from the cell body). The connecting and supporting cells are called glial cells. One type of glial cell is the Schwann cell which envelopes the peripheral nerves and produces myelin sheaths which insulate the nerve axons. Examples of nervous tissue cells are shown in Figure 4.4d.

Blood and lymph tissue

This tissue type includes all cells in the peripheral blood and their precursor cells in the bone marrow and lymph glands. Some of these cells exchange oxygen and carbon dioxide with the external environment, whereas others facilitate the immune response. Examples of blood cells are shown in Figure 4.4e.

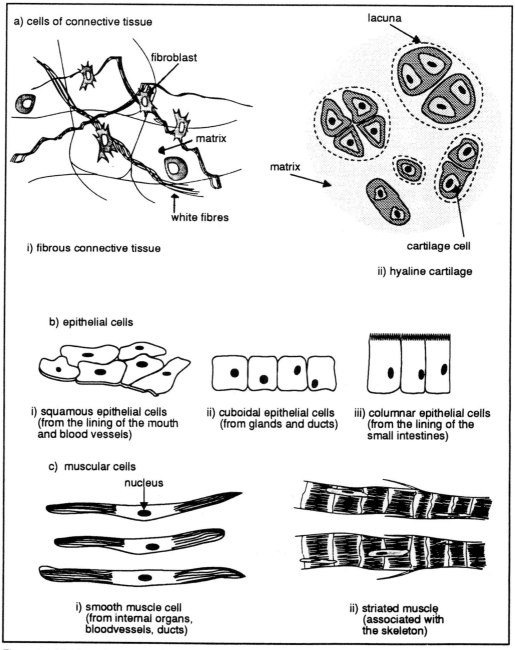

Figure 4.4 Highly stylised representations of the principal cell types in vertebrate tissues.

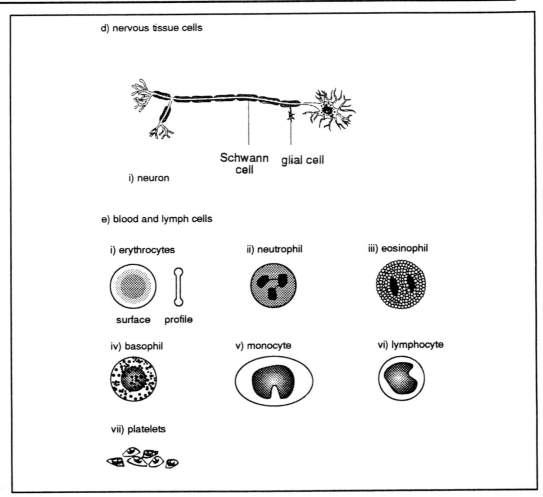

Figure 4.4 Continued

4.4.2 Cell division and differentiation *in vivo*

All of the different types of tissue mentioned above develop from a single cell, the fertilised ovum, by a process of cell division and cell differentiation (see Figure 4.5). The ovum itself is totipotent, which means it is not irreversibly committed to any developmental fate and is able to give rise to any cell type.

totipotency

At each stage of development, cell division and cell differentiation are carefully regulated. In a young animal, cell multiplication exceeds cell death, so that the animal increases in size. In the adult, cell division and cell death are balanced to produce a steady state. Cells multiply through the processes of mitosis and cytokinesis linked together in the cell cycle. We will assume that you are familiar with these fundamental processes. A general description of the cell cycle is given in the BIOTOL book 'Infrastructure and Activity of Cells'.

mitosis
cytokinesis

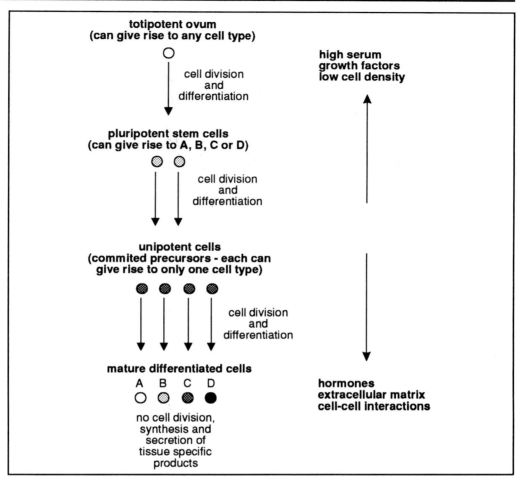

Figure 4.5 Schematic diagram to show the different stages in the development of an organism. (See text for discussion).

differentiated
cells
commitment
pluripotent cells

Most of the cells of the adult organism are quiescent, although some adult cell types do divide rapidly, for example the intestinal cells have a half life of a few days before they die and are replaced. The quiescent cells of the adult animal are generally differentiated in other words they express the phenotypic properties characteristic of the functionally mature cell *in vivo*. Even before a cell actually differentiates, its destiny is fixed in a process called commitment. However, not all cells are committed to a specific lineage. Many cells in the adult animal remain pluripotent, that is not irreversibly committed to a single development fate but able to give rise to only a limited number of different cell types. In the haematopoietic system a small population of pluripotent undifferentiated stem cells typically give rise to committed precursor cells which progress towards terminally differentiated cells, loosing the capacity to divide as they reach the terminal stages. The fully mature terminally differentiated cell will not normally divide.

Once the cell has progressed down a particular lineage to a point at which the mature phenotype is fully expressed and beyond which the cell cannot proceed, it is said to be terminally differentiated.

de-differentiation   Differentiation is not necessarily irreversible. Some differentiated cells can de-differentiate thus losing their differentiated properties and resume proliferation. This happens when cells become malignant and when they are grown in culture. These cells loose at least some of the specific phenotypic properties associated with the mature cell, but again this loss is not necessarily irreversible.

∏   Describe a role for de-differentiation in the adult organism.

De-differentiation is important in tissues where occasional renewal of cells is required, for example, during wound repair in the skin. Some of the already differentiated cells may loose some of their differentiated properties and re-enter the cell cycle. The cells proliferate until the tissue has gained the appropriate cell density, at which point proliferation stops and differentiation is re-induced.

In the adult organism, commitment, differentiation, de-differentiation and terminal differentiation all play a role.

4.4.3 Factors needed for differentiation *in vivo*

organogenesis and cell-cell interactions   Interactions between different types of cells are responsible for initiating and promoting differentiation during normal development. This is apparent during organogenesis, when interactions between epithelial precursors and stromal precursors induce epithelial differentiation, but such cell-cell interactions also play a role during differentiation in the adult organism. The exact nature of these interactions has been more difficult to define and the inducing signals that pass between cells are often not known.

Here we will briefly describe the roles of hormones, paracrine factors, the extracellular matrix and cell-cell contact in differentiation *in vivo*. We will not give specific details at this stage as the study of these factors is more appropriate to the study of cell biology and physiology. Aspects of cell differentiation are discussed in the BIOTOL texts 'Infrastructure and Activities of Cells' and 'Functional Physiology'. Our knowledge of cell differentiation and development in mammals is particularly well developed in our understanding of the development and function of the cells involved in the immune system. These are described in the BIOTOL text 'Cellular Interactions and Immunology'.

Hormones

One obvious way in which cells can interact is through the production by one cell of intercellular signalling molecules, such as hormones, which have their effect on other cells. Strictly speaking the term hormone refers only to substances secreted by the endocrine glands directly into the bloodstream, and there are many other types of intercellular signalling molecules.

Paracrine factors

Paracrine factors are molecules which do not travel via the bloodstream, but have their effect on target cells which are close to the signalling cell. Paracrine signalling plays a vital role in development and differentiation *in vivo*.

Extracellular matrix

laminin and
fibronectin
tenascin

The extracellular matrix is a mixture of collagens, non-collagenous proteins (such as laminin and fibronectin) and carbohydrate-rich molecules. Another way in which cell differentiation is induced is by modification of the extracellular matrix, for example tenascin (an extracellular matrix protein) is synthesised by stromal cells and is thought to participate in inducing differentiation of epithelial tissues *in vivo*. In the intact organism, the extracellular matrix forms the natural growth substratum and often plays an essential role in the induction of differentiation.

Cell-cell contact

gap junctions

Finally, cells interact directly, for example via gap junctions. These are large channels formed by rings of hexagonal protein subunits called connexons, which can be opened and closed. They are present in most eukaryotic membranes and serve as a passageway between the interiors of neighbouring cells, allowing ions and small molecules to flow between the cells. Gap junctions are important in intercellular communication, and this is particularly clear in excitable tissue such as heart muscle where cells are coupled by the rapid flow of ions through these junctions, thus assuring a rapid and synchronous response to stimuli. Communicating channels like gap junctions are also essential in development and differentiation. Cells also interact via surface information exchange. The molecules expressed on the surface of one cell can directly affect the behaviour of an adjacent cell.

In summary, soluble factors (hormones and paracrine factors), changes in the composition of the extracellular matrix and direct cell interactions all play a role in development *in vivo*.

4.4.4 Animal cell types grown *in vitro*

We mentioned in the first chapter of this text the types of cells that have been grown in culture to date.

∏ See if you can remember the different types of cells which can be grown in culture by making a list of them.

The different cell types which can grow in culture include:

fibroblasts
epithelial tissue (liver, lung, breast, skin, bladder and kidney) cells
neural cells (glial cells and neurons, although neurons do not proliferate *in vitro*)
skeletal cells, cardiac cells and smooth muscle cells
skeletal tissue (bone cells and cartilage cells)
endocrine cells (adrenal cells, pituitary cells, pancreatic islet cells)
melanocytes
many different types of tumour cells.

You might anticipate that these cell types have specific requirements to be cultured *in vitro*.

4.4.5 The differentiated state *in vitro*

If differentiated cells are to be studied in culture, then it is important to confirm that these cells still have their differentiated characteristics when they are grown *in vitro*. Many cultured cells continue to perform their specialised functions *in vitro*. The vast majority of published studies using tissue culture describe fibroblasts, since these cells grow rapidly in culture.

∏ Recall the different types of animal tissue which are found *in vivo*.

In section 4.4.1, we said that animal tissues are classified as epithelial, connective, muscular, nervous and blood and lymph. We will now look at how each of these tissue types grow *in vitro*.

Connective tissue

Cultured fibroblasts have the morphology of tissue fibroblasts although they tend to be less well differentiated and appear to be precursors of fibroblasts. With appropriate stimulations these cells can differentiate in culture into many cell types, such as adipose tissue and connective tissue cells. In many studies fibroblasts are treated as convenient prototypical cells for study.

Epithelial tissue

Many studies have been conducted using cultured epithelial cells. Epithelial cells, like fibroblasts, are grown on plastic dishes to which they adhere tightly due to their secretion of extracellular matrix proteins such as laminin. However, early studies of epithelial cells were disappointing, since the important characteristics of these cells were lost in culture. It was known that the level of expression of many differentiated products is under hormonal control, so cells in culture were supplemented with specific hormones. However, fully differentiated epithelial cells still lost their tissue-specific function within 3 days in culture, even in the presence of the appropriate soluble factors and hormones. The reason for this now appears to be that the establishment of correct cell polarity and cell shape is particularly important in epithelial cells in culture.

use of suitable substrate or co-culture may maintain required differentiated status

De-differentiated cells could be returned to their normal state in many cases by growing them on a characteristic substrate, or in the presence of another cell type. For example, mammary epithelial cells will show normal ductal morphogenesis when grown on a fat pad or on collagen type I gels. They are able to express and secrete milk proteins when the appropriate hormones are added. The culture of mammary epithelial cells demonstrates the importance of considering cell-cell and cell-extracellular matrix interactions, as well as soluble factors which induce differentiation.

∏ See if you can devise culture conditions which will enhance the probability that skin cells (epithelial cells) will differentiate into cornified or keratinocytes *in vitro*.

We hoped you would use the following logic. Skin epithelial cells normally grow and differentiate as outer-layers surrounding other cells. Thus if we try to culture these cells as layers on top of other cells we may mimic their natural (*in vivo*) conditions. In fact, skin cells in culture can be induced to differentiate by adding certain growth factors (such as epithelial growth factor) and growing the cells on 'feeder' layers of other cell types. The skin cells will divide and differentiate into cornified cells or keratinocytes

'feeder' layers

that make up the outer surface of the skin. Culturing cells on top of a feeder layer of cells can enhance growth and/or differentiation of those cells. It is not always entirely clear why this works but it may be that the feeder layer is providing paracrine factors, extracellular matrix, or cell surface molecules which provide the appropriate interactions necessary for the differentiation of the other cell type.

Muscular tissue

Permanent lines of muscle precursor cells called myoblasts have been developed and when these cells reach a certain density, they can carry out functions which are similar to the functions of muscle cells *in vivo*. They fuse with each other and form cross-striated multi-nucleated muscle fibres (myotubes). A genetically homogeneous clone of rat myoblast cells, called L6, has been isolated and used to analyse the biochemical changes which take place during differentiation *in vitro*.

myoblasts L6 cells

Nervous tissue

Some of the earliest tissue culture experiments were performed using neurons. In 1907, Ross Harrison cultured pieces of tissue from the medullary tube region of frog embryos in clots of frog lymph. The neural cells maintained their typical neural morphology *in vitro* and axons grew out from some of the cells.

Tissue culture has been particularly useful in studying the factors that regulate development of the peripheral nervous system. At first the heterogeneity of neural and glial cells which constitute the nervous system proved to be a barrier to understanding the physiology and biochemistry of individual subclasses of neurons and glial cells. However, avian and mammalian peripheral neurons are conveniently located in discrete ganglia and each ganglion is comprised of a single neuronal type with associated glial cells. These ganglia are easily dissected from avian and mammalian embryos or neonates (1-48 hours old). Neonatal mouse dorsal root ganglia from the spinal cord have recently been used to obtain pure cultures of Schwann cells and pure cultures of neurons. The protocol for separating these two cell types depended on differences in the ability of these cells to adhere to certain substrates. Once isolated, the pure cultures have been used to study the effects of different growth factors on neural differentiation.

Blood and lymph

leukocyte progenitor cells

Isolation and culture of human peripheral blood leukocytes became standard practise in the 1960s and many of the precursors of leukocytes (lymphocytes, monocytes and macrophages) can carry out at least part of their differentiated function in culture. The clonal growth of these progenitor cells in culture was the key to sorting out the haemopoietic (haematopoietic) lineages and the discovery of the polypeptide hormones which regulate haemopoiesis.

Erythroleukaemia cells (abnormal precursors of red blood cells) can be induced to produce haemoglobin and undergo structural changes in the cell membrane associated with normal red blood cell maturation. However, many specialised cell types cannot be cultured, even when cell-cell and cell-extracellular matrix interactions and soluble factors are taken into account.

Name one reason why it may be difficult to culture some cells, even given the correct environment.

Many cell types stop dividing at a specific stage in development *in vivo*, and so it is not surprising that they cannot be induced to divide *in vitro* (see Figure 4.6). For example, there is a steady loss of brain cells in adults with little or no replacement. An adult brain cell would, therefore, be unlikely to proliferate *in vitro* since it does not do so *in vivo*. This does not mean that it will be impossible to find conditions in which adult brain cells can be induced to divide *in vitro*, but it will certainly involve a lot more research than for many other types of cells.

4.4.6 Characterisation of cell types *in vitro*

A number of different techniques are used here depending on the tissue type.

Morphology

The simplest way to recognise cell types is to examine their morphology under the microscope. Epithelial cells tend to be regular, polygonal, with a clearly defined edge. Fibroblasts tend to be bipolar and form characteristic parallel arrays and whorls visible to the naked eye. These characteristic morphologies apply to confluent adherent cells and different morphologies are seen at lower confluence. In addition, morphology may change with the substrate on which cells are cultured and with the type of medium used. So comparative observations should always be made at the same stage of growth, at the same cell density, in the same medium and growing on the same substrate.

The terms epithelial and fibroblastic are often based on a description of the appearance rather than the origin of cells although officially the terms 'fibroblast-like' or 'fibroblastoid', and 'epithelial-like' or 'epitheleoid' should be used until the identity of the cell has been authenticated.

Apart from analysis of morphology there are other techniques to identify cell types, but before we move on to look at these test your knowledge of the techniques referred to so far in this chapter. Attempt SAQ 4.3 and read our response before continuing with the text.

SAQ 4.3

1) What are the techniques used to verify (a) the species and (b) the individual from which a cell line originated?

2) Which of the techniques you have used in answer to 1b) could also be used to characterise cell types, and why?

For more specific identification of cell type, there are a number of markers which can be studied. These include:

- intermediate filament proteins;

- cell surface antigens;

- expression of functional properties of the mature cell;

- analysis of isoenzymes.

We will examine each in turn.

Intermediate filament proteins

cytokeratins in
epithelial cells,
vimentin in
stromal cells,
desmin in
myogenic cells,
neurofilaments
in neurons

Eukaryotic cells contain a network of proteins collectively called the cytoskeleton which maintain cellular morphology and also play a role in intracellular transport, cell motility, mitosis and meiosis. Cytoplasmic intermediate filament proteins form an essential part of this cytoskeleton and are frequently used in the identification of cell types. For example cytokeratins are indicative of epithelial cells, vimentin of stromal cells, fibronectin of fibroblasts, desmin of myogenic cells, and neurofilaments of neurons and glial cells. The technique used is to enrich cell extracts for intermediate filament proteins and then to use antibodies against various intermediate filament proteins to identify the cell type. Alternatively, immunofluorescence can be used to directly stain the cells.

Cell surface antigens

Cell surface antigens are particularly useful in sorting subspecies of lymphocytes. This area has been expanded in recent years due to the availability of monoclonal antibodies.

Expression of functional properties of the mature cells

Specialised functions *in vivo* are often expressed in the activity of specific protein products, for example erythrocytes express haemoglobin and mammary epithelial cells express milk proteins. Induction of these products often requires specialised conditions, such as hormone additions and specific substrates.

Analysis of isoenzymes

Another important technique is the analysis of isoenzymes. Some cell lines do not express any cell-type specific enzymes, but the pattern of isoenzymes for some housekeeping enzymes may be characteristic for that cell type. The methodology has been described in 4.2.2. In addition to the three enzymes mentioned in section 4.2.2 there are a number of enzymes which can be used for characterisation of human cell lines including phosphoglucomutases (PGM 1 and 3), 6-phosphogluconate dehydrogenase (PGD), malic enzyme (ME-2) and adenylate kinase (AK-1). A single example can be used to illustrate how this can be applied.

Mammals produce two types of polypeptides which have lactate dehydrogenase activity. One, which is characteristic of heart tissues, we will label as H, the other, which is typical of muscle, which we label as M. These polypeptides associate to form tetramers and it is as a tetramer that the enzyme is found in cells.

∏ If a cell produces both types of polypeptide chains, how many different forms of the enzyme might be found in a cell?

The answer is 5. They are H_4, H_3M, H_2M_2, H_1M_3 and M_4. However, the proportions of each of these are different in different cell types. Since the different forms have different electrophoretic mobilities, it is easy to separate these isoenzymes to give a distinctive pattern. Cell lines derived from different cell types can be descriminated in this way.

| **SAQ 4.4** | A number of newly established cell lines were characterised as to their tissue of origin. Below we have provided two lists. One lists types of cells, the other lists a series of properties. Assign the properties to the appropriate cell type(s). Not all cell types will be relevant and some may be used more than once. |

Cell Types

a) Myoblasts

b) Human mammary epithelial cells.

c) Mouse embryo fibroblasts.

d) Mammary fibroblasts.

e) Erythrocytes.

Properties

1) Express the milk protein casein.

2) Have a bipolar morphology.

3) Express haemoglobin.

4) Express fibronectin.

5) Express a cytokeratin.

6) Express desmin.

7) Have a regular, polygonal morphology.

The use of cell type specific markers makes it possible to determine the lineage from which cultures are derived but does not necessarily indicate the position of cells within this lineage. It is essential not only to characterise the lineage of cells being used, but also the stage of differentiation.

4.4.7 Identifying the stage of differentiation

cells in culture usually do not express differentiated characteristics fully

In tissue culture, one of the starting points is that we want cells to divide in culture. For cells to proliferate it is likely that they represent a precursor cell rather than a fully differentiated cell which would not normally proliferate (closer inspection of Figure 4.5 shows that mature differentiated cells do not usually divide). Since cells *in vivo* do not have the ability to divide and express some differentiated products at the same time, it is not surprising that many cell lines (which still have the ability to divide) do not fully express differentiated properties. Sometimes, a culture contains cells at different stages of differentiation: stem cells, precursor cells and differentiated cells. Such a culture is phenotypically heterogeneous although it is genetically homogeneous. Other cultures are more uniform.

There have been attempts to re-induce the differentiated phenotype in pure populations of cells by re-creating the correct environment. Since different conditions are required for propagation and differentiation, these experiments usually involve first growing up a large number of cells and then stopping cell division and inducing differentiation by adding the appropriate inducers, or by creating the correct cell-cell or cell-extracellular matrix interactions. New developments in cell culture have focused on the refinement of culture conditions by defining more precisely the substratum and growth factors required for an individual cell type. We will examine this in a little more detail in the next sub-section.

4.4.8 Re-inducing the differentiated state *in vitro*

One of the ultimate aims of cell culture study is to design the cell culture environment to manipulate the differentiated state *in vitro*. The challenge has been to determine the factors that delay, direct and promote different lineages of cells towards terminal differentiation. Here we examine three of the main factors which promote differentiation; the extracellular matrix, cell-cell interactions and soluble inducers.

Extracellular matrix

coating of
vessels with
collagen

In the intact organism, the extracellular matrix forms the natural growth substratum and until the late 1970s the extracellular matrix was considered to be a static structural support for cells *in vivo*. However, recently studies with cell cultures have revealed that the extracellular matrix consists of distinct molecules that affect cell behaviour. We have already mentioned the role of some of these components in inducing differentiation, for example in section 4.4.2 we discussed the role of tenascin in inducing differentiation of epithelial tissue *in vivo* and in section 4.4.4 we discussed the role of collagen I in the ductal morphogenesis of mammary epithelial cells. This led to idea that it may be possible to influence the differentiated state by varying the components of the extracellular matrix *in vitro*. Artificial matrices are often used to simulate the environment *in vitro*, for example collagen is often used for coating tissue culture dishes, and is particularly important for the functional expression of many epithelial cells.

Cell-cell interactions

In addition, direct cell-cell interactions are very important. Homologous cell interactions occur optimally at high cell density and this may allow gap junctional communication and surface information exchange. Heterologous cell interactions are often responsible for initiating and promoting differentiation during normal developments. Since cell-cell interactions are so important in differentiation, many people have reverted to organ culture or attempted to re-create tissue-like structures *in vitro* by re-aggregating different cell types (see Chapter 9).

Soluble inducers

Physiological inducers of differentiation include hormones, paracrine factors, vitamins (such as vitamin A), and inorganic ions (such as Ca^{2+}). In addition a number of non-physiological compounds have been found to induce differentiation *in vitro*, for example DMSO induces differentiation of certain cell lines.

Given the correct environmental conditions, partial or complete differentiation is achievable in cell culture. We will return to this point later.

4.5 Microbial contaminations

We mentioned in Chapter 1 that the development of animal tissue culture techniques was to a large extent dependent upon the ability to keep cells free of contamination. Thus preventative methods were developed such as aseptic technique, addition of antibiotics and the use of laminar air flow cabinets. Nevertheless, contaminations do still occur and it is important to recognise these when they arise.

EC-guidelines
and Directives

The detection of contaminating micro-organisms is also essential as a quality control measure. In cases where the cell culture is used to produce a product which is to be marketed (for example as a diagnostic or therapeutic reagent), market authorisation is only given if the quality of the product can be assured. In the European Community, the EC Commission has published an extensive series of Guidelines and Directives covering this field. Although it is beyond the scope of the present volume to deal with this subject extensively, it is important that you are aware that such guidelines exist. These guidelines are published in the Official Journal of the EC (some examples are given in the BIOTOL texts 'Biotechnological Innovations in Health Care' and a list is given in 'A Compendium of Biotechnological Practices: Safety, Good Practice and Regulatory Affairs'). In the USA, similar guidelines are laid down by the US Food and Drug Administration.

Here we will divide our discussion into:

- bacteria, fungi and yeasts;

- mycoplasmas;

- viruses.

4.5.1 Bacteria, fungi and yeast

Bacteria, fungi and yeast are fairly easy to identify and signs of contamination include rapid changes in pH, cloudiness in the medium, extracellular granularity under the microscope, and unidentified material floating in the medium. When these symptoms appear, cultures are generally discarded, especially if frozen stocks are available.

The organisms commonly found as contaminants in cell culture are listed in Table 4.1.

Bacteria	Fungi/yeasts
Pseudomonas spp	Aspergillus spp
Staphylococcus spp	Candida spp
Escherichia coli	Penicillum spp

Table 4.1 Organisms commonly found in animal cell cultures.

Cell cultures may be examined for these contaminants by sub-culture into suitable media. The media most commonly used for this purpose are; thioglycollate medium, soybean-casein digest and tryptone soya broth. Usually a small sample (1 ml) of the culture is inoculated, in duplicate, into one or more of these media and incubated at 26°C and 37°C. These cultures are examined daily over two weeks for signs of growth. These tests are, of course, designed to detect low levels of contamination in the original animal cell culture.

use of antibiotics

It is preferable to grow animal cells in antibiotic-free media as the presence of antibiotics may suppress but not eradicate antibiotic-resistant contaminants. If, however, a valuable culture becomes contaminated, then a short-term exposure to antibiotics may enable recovery of the culture. If contamination by a Gram positive organism is suspected penicillin is used, whilst streptomycin is usually effective against Gram negative organisms. Fungal contamination is usually treated with amphotericin B or hystatin. However, preferably such cultures should be discarded.

There are some general recommendations that may be offered to reduce the chances of microbial infection. We have given these in Table 4.2.

Only use cells from a reliable source.

Carry out quality control tests on receipt of a culture and regularly during sub-culturing.

Carry out quality control tests on media and other reagents before using them.

Use good microbiological practices (eg disinfect bench tops, wash hands regularly).

Do not use the same media samples for different cell lines.

Do not use antibiotics unless contamination is suspected.

Table 4.2 Recommendations to reduce microbial infection.

4.5.2 Mycoplasma

Contamination by mycoplasma is more serious since it is not easily visible. Mycoplasma are small, self-replicating organisms, which often proliferate without any overt changes in the contaminated cell line. They can be as small as 0.3 µm, and they are not visible although they can seriously affect cellular biochemistry, antigenicity and growth characteristics.

We must also include members of the genus *Acholesplasma*. They are similar to other members of the order Mycoplasmatales except they do not depend upon steroids to grow. All other genera of this order have an absolute requirement for steroids. The common mycoplasmal infections are strains belonging to just five species. These are largely host specific and are:

M.orale	human
M.fermentaris	human
M.arginini	bovine
Acholesplasma laidlavii	bovine
M.hyorhinis	porcine

staining for DNA to detect mycoplasma

Cultures may become infected with mycoplasma from media, sera, trypsin or the operator. It is, therefore, important to test for mycoplasma at regular intervals (every 1-3 months). The protocols for detecting mycoplasma contamination are simple and reliable. Mycoplasma DNA binds to a fluorescent dye (fluorochrome dye, Hoechst 33258). Under the microscope, cells stained with this dye have fluorescent nuclei. Also extranuclear DNA (in plastids and mitochondria) and mycoplasmal DNA fluoresces. The advantage of this technique is that it can be used to detect non-cultivatable mycoplasma strains.

∏ Suggest two types of control.

We hoped you would have included a control in which no test cells were added and also a control in which a known mycoplasma is deliberately added. The first of these is a negative control, the second is a positive control. Good strains to use for positive controls are *M.hyorhinis* and *M.orale*. Some researchers do not include a positive control, they would prefer to keep mycoplasma cultures well away from the animal cell culture laboratory!

The test samples are incubated for 8-12 hours and fixed with methanol:acetic acid (v/v 3:1), air dried and strained with Hoechst 33258. Examination is done using UV incident light. The presence of small fluorescent filaments or cocci are indicative of mycoplasma.

∏ Describe two circumstances in which the test would be invalid.

If similar filaments or cocci were present in the negative control or were absent from the positive control, this would indicate that the test was not working properly and would invalidate the results.

Recently, two new types of techniques have been introduced. Bethesda Research Laboratories market a system called MycoTect. In this test, cells are incubated with 6-methyl purine deoxyriboside. This is non-toxic to mammalian cells. Mycoplasmal cells, however, metabolise this deoxyriboside to form either 6-methyl purine or 6-methyl purine riboside. These are toxic to mammalian cells. Thus mammalian cells harbouring mycoplasma are killed.

In contrast, Gene-Probe market a system called 'Gene-Probe'. This consists of ^{3}H-labelled DNA probe which will hybridise to mycoplasmal DNA but not to mammalian DNA. Hybridised and free ^{3}H-DNA probe are easily separated electrophoretically.

| SAQ 4.5 | The following data were obtained using the Gene-Probe system for detecting mycoplasma in cell culture. |

	Control 1 (no added animals cells)	Control 2 (mycoplasma added)	Culture A	Culture B	Culture C
% radioactivity hybridised to DNA*	5% (4.6-5.4)	78.5% (76.5-80.5)	6% (4.6-7.4)	11.5% (11.4-11.6)	7.1% (6.6-7.6)

* Results are reported as means of two duplicate samples (actual results are given in brackets).

From this data, decide if cultures A, B and C are contaminated by mycoplasma and explain any further tests you would carry out.

4.5.3 Viruses

Infection of cell lines with viruses can be very problematic, again, because they often remain undetected. Persistent or latent infections may exist in cell lines and not be noticed until immunological, cytological, ultrastructural or biochemical tests are applied. In spite of the variety of screening procedures for viruses which exist, latent viruses which do not react in these tests will remain undetected, presenting a possible health hazard to the operator and also affecting the properties of the cultured cell.

immunological
assays

The most commonly employed technique is to use immunological assays. The problem is to decide which viruses to assay for. Since there is such a wide variety of viruses which may infect cells and because some exhibit extensive 'immunological drift' (that is different strains are antigenically distinct and new antigenic types arise from existing strains quite rapidly) it becomes virtually impossible to declare that a cell culture is 'virus-free'. All we can declare is that the culture has been tested for particular viruses. Nevertheless, we can apply some logic to our testing. For example, if we were cultivating mouse myeloma cells we would probably confine our testing to mouse viruses especially retroviruses.

Let us give specific example. Cilag Bendux market a monoclonal antibody called orthoclone OKT3, a therapeutic antibody used as an immunosupressor. To gain market authorisation for this monoclonal antibody produced in murine cells, tests were carried out for a range of murine leukaemia viruses (eg $S^{-1} L^{-1}$, XC plaque, Xenotropic Murine Leukaemia viruses, Ecotrophic Murine Leukaemia viruses). In addition, assays designed to detect other murine viruses (eg Reovirus 3, K virus, LEM virus, mouse hepatitis virus, mouse adenovirus) were also conducted.

Summary and objectives

In this chapter we have looked at ways of characterising cell lines, with respect to the species and individuals from which they originated, the tissue of origin, the stage of differentiation and the possible presence of microbial contaminants. Such analyses are essential for all uses of animal tissue culture cells. The information on differentiation *in vitro* has been presented within the context of our knowledge of events *in vivo*. The micro-environment (hormones, paracrine factors, extracellular matrix, adjacent cells) is the key determinant of the direction of the cell in terms of cell division and differentiation. Many differentiated cells need extracellular matrix components and the correct hormonal environment to express differentiated characteristics *in vivo*. Thus attempts have been made to simulate these conditions *in vitro*.

Now that you have completed this chapter you should be able to:

- describe the approaches used for species identification (cytogenetics, isoenzymology and immunological tests);

- specify the methods used to analyse intra-species contamination (cytogenetics and recombinant DNA methodology, high resolution 2-D electrophoresis, allozyme analysis and antigen analysis);

- select appropriate techniques to detect inter- and intra-species contamination of animal cell cultures;

- describe the approaches used in identification of cell type (cell surface antigens, intermediate filament proteins, differentiated products);

- justify the measures taken to induce the differentiated stage *in vitro* by analogy to the events which occur during differentiation *in vivo*;

- select suitable procedures for detecting microbial contamination of animal cell cultures;

- describe the techniques used to detect mycoplasma and viruses in animal cell cultures;

- explain why the detection of mycoplasma and viruses is important in quality assurance in production processes involving animal cells.

Preservation of animal cell lines

5.1 Introduction   100

5.2 Variation and instability in cell lines   101

5.3 Preservation of cell lines   102

5.4 Freezing down   105

5.5 Thawing frozen cells   109

5.6 Quantitation of cell viability   109

5.7 Cell banks   110

Summary and objectives   114

Preservation of animal cell lines

5.1 Introduction

the evolution of
cell lines
Thousands of cell lines derived from human and animal sources are in existence today. Some derived from normal tissues have a limited growth potential, whilst others, which have been derived from tumours, or transformed to behave as tumour cells, can be propagated in continuous culture for many years.

As a cell line evolves from primary culture to form a secondary or established cell line there is a danger of introducing genetic or phenotypic instability or contamination with other cells, bacteria, fungi or mycoplasma.

It is essential to use cell lines which are well defined and free from contamination in order for experimental results to be valid. For this reason it is vital that laboratories using animal cell lines should:

- standardise culture conditions;

- select a period in the life of the cell line where variation is at a minimum;

- select a pure clone, or well characterised cell line, if possible;

- preserve a seed stock of the cell line which can be used to replenish the working stock at intervals.

Maintaining cultures simply by sub-culturing is expensive in terms of time and materials. This system is also susceptible to contamination. Every time we open a vessel to inoculate it, there is a finite chance of contaminant micro-organisms gaining access to the culture. These problems are particularly acute when cultures are needed to be maintained for use over many years. This chapter examines the techniques available for long term storage of cell lines.

From our discussions in earlier chapters, you should be familiar with cloning and characterisation of cell lines. In this chapter, we will first re-examine how variations and instability arise in cell lines and then examine how desirable cell lines may be preserved. We will provide a brief historical background before considering the selection of cell lines for storage. We will also examine the stages in cell preservations. We will include a discussion of the damage that may be inflicted on cells during their preservation and then explain how cells may be resuscitated. In the final part of the chapter, we will describe the availability of cell lines maintained in cell banks.

5.2 Variation and instability in cell lines

5.2.1 Culture conditions

use of batch
tested media

Once the optimal conditions for maintaining a cell line have been established (see Chapter 3), these should be adhered to as far as possible. Any changes in the constituents of the cell culture medium, batch differences in serum, differences in temperature and gases can alter the growth characteristics of the cell line. Laboratories which regularly work with animal cell lines usually buy batch-tested medium and serum in large quantities often sufficient to last around a year.

5.2.2 Selective overgrowth can rapidly alter a cell line

We learnt earlier that in a primary culture there will be a number of different cell types growing at different rates. Some may grow very quickly and overgrow the more slowly dividing or non-dividing cells. For example, if cells are explanted from a colon carcinoma the first culture will contain cells which are mainly epithelial. After sub-culture, however, these epithelial cells diminish in number and the fibroblast cells take over. Eventually all the epithelial cells disappear and the fibroblasts form a secondary cell line.

seed and
working stocks

If, for a particular experiment, the epithelial cells need to be preserved then it will be necessary to produce a clone of the cells required and also to freeze down some of the primary culture in case disaster strikes and a clone cannot be produced. Once the required clone is obtained, it is vital to freeze down a seed stock which can be used to replenish the working stocks should these get contaminated or lose the particular characteristics required.

5.2.3 A transformed cell can appear in an established cell line

Sometimes a cell can become transformed during the life history of an established or semi-established cell line and take over very rapidly. The transformation event may enable it to have a much greater growth capacity than the slower growing parent cell line. (We will examine this in greater detail in Chapter 6).

5.2.4 Genetic instability can alter the phenotype of a cell line

Established cell lines have a high incidence of genetic variations partly because the rate of spontaneous mutations is higher in cells which are proliferating rapidly and partly also because there is no mechanism *in vitro* for the elimination of mutants as they appear. These genetic changes can be expressed as phenotype changes and thus the characteristics of the cell line may change altogether. Cells which are to be preserved should be genetically characterised (see Chapter 4).

5.2.5 Senescence

A finite fibroblastoid cell line such as that derived from human embryo lungs will be stable for only a limited number of generations. The cell line will become stable at around the fifth or sixth passage (generation) and can be used up to the 30th generation. After this senescence can become a problem and cells may lose certain receptors as well as the ability to replicate.

By preserving some of the early sub-cultures, it is possible to extend the time in which work can be conducted on the cell line. We can represent this situation in the following way.

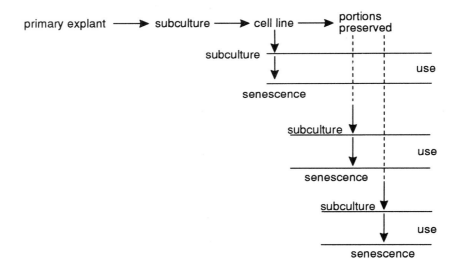

5.3 Preservation of cell lines

5.3.1 Historical background

early studies
with
spermatozoa

In 1776 Spallanzi published one of the earliest observations of the effect of 'cold' on cells. He noticed that spermatozoa chilled with snow become motionless and that this loss of activity was totally reversed upon increase in temperature. A hundred years later Mantegazza (1866) experimented on the preservation of spermatozoa and noted that cells show a good degree of resistance to freezing and thawing. This work had relevance in animal husbandry, but it was not until the 1940s that work on the preservation of cells other than spermatozoa really took off. In 1945 Parkes showed that the survival of stored cells was increased by the presence of glycerol and by increasing the cell density. Around this time other workers noted that the major cause of cell damage could be attributed to the rise in concentration of electrolytes as water was removed into ice crystals.

| SAQ 5.1 | Describe the likely biochemical and mechanical changes that occur within the cell and its immediate environment as the temperature is reduced from 37°C to -20°C. |

cryopreservation   The preservation of animal cells has become dependent upon using low temperature to render cells metabolically inert. The term cryopreservation is given to this method of cell storage. As we shall see, for optimum results storage is usually done at liquid nitrogen temperatures (-196°C) although temperatures of -70°C to -80°C may be sufficient to maintain cell viability for a few months.

An overview of the stages used in the preservation of animal cell lines is shown in Figure 5.1. An outline of the stages in selecting cells for storage is shown in Figure 5.2.

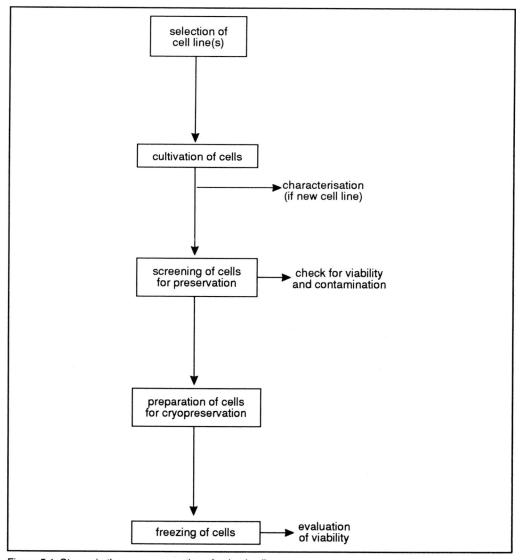

Figure 5.1 Stages in the cryopreservation of animal cells.

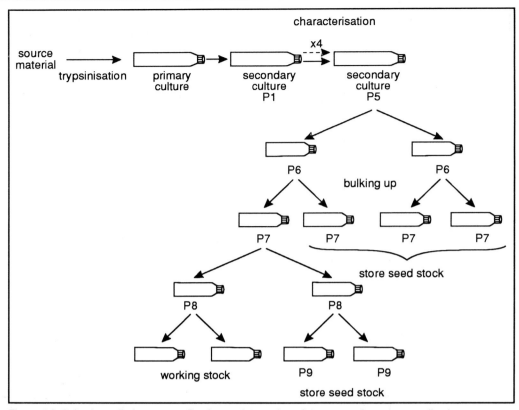

Figure 5.2 Selecting cells for storage. P refers to the number of passages after primary cell culture.

Cells preserved in the late log phase are more resistant to the traumas of freezing and thawing and will continue to replicate once thawed.

It is usual to test the viability of cells before cryopreservation. This can be done using a staining procedure which enables a distinction to be made between viable and dead cells. Trypan blue is commonly used. It will stain dead cells but is excluded from live cells. Usually viabilities greater than 95% are required and cultures showing lower viabilities are usually discarded.

∏ What would you also check for in cultures of cells which are to be used for cryopreservation?

You should also check that the culture is free of microbial (especially mycoplasma) contamination. The tests for this type of contamination were discussed in Chapter 4.

Cells to be used for storage are collected using the same methods as used for their routine sub-culture. For example, attached cells are removed using trypsin whilst cells in suspension may be collected by gentle centrifugation. Collection must, of course, be done aseptically.

∏ Give two reasons why cells in suspension are collected by gentle centrifugation.

The two main reasons are that centrifugation reduces the volume that needs to be frozen which enables the freezing process to be more controlled. Secondly, gentle centrifugation is used to minimise the extent of mechanical damage to the cellular membranes.

<table>
<tr><td>

SAQ 5.2

</td><td>

A culture has been characterised and gave the following results.

1) Cell density = $1 \times 10^6 \, \text{ml}^{-1}$.

2) Free of bacteria.

3) Free of mycoplasma.

4) Free of fungi

5) 85% viable (as tested using trypan blue).

Is the culture suitable for cryopreservation?

</td></tr>
</table>

5.4 Freezing down

5.4.1 Temperature of storage

storage in
liquid nitrogen

Cells can be stored for varying lengths of time at temperatures between -70°C to -196°C. For short term storage, for example for no more than a few weeks, cells can be preserved at -70°C in standard mechanical freezers. For longer term storage, it is necessary to use much lower temperatures and liquid nitrogen is the medium of choice. Cell banks and most reasonably sized laboratories routinely preserve cells and other valuable material in liquid nitrogen containers. The containers are basically robust, heavily insulated, vessels into which liquid nitrogen is poured at regular intervals in order to maintain the required temperature. Cells can be stored in either the vapour phase (-120°C) or the liquid phase (-196°C) of liquid nitrogen (see Figure 5.3). Storage at -120°C does not significantly reduce cell viability, although for very precious material or for indefinite storage the liquid phase is often preferred.

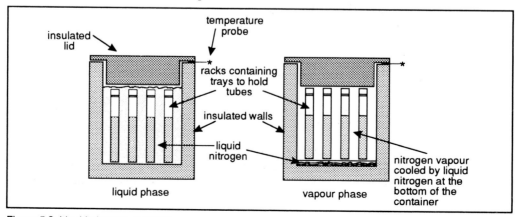

Figure 5.3 Liquid nitrogen containers.

advantages of
using N₂
vapour

There are advantages and disadvantages associated with both systems. Storage in the vapour phase is generally safer: seepage of liquid nitrogen into the storage tubes (ampoules or cryotubes) does not occur so the risk of explosion is reduced. If liquid nitrogen enters the ampoule, the sudden evaporation of the liquid on removal of the ampoule from the cryostat may cause the ampoule to explode violently. The risk of the contents of the tubes, which may contain infectious material, leaking out of the tube and contaminating the liquid nitrogen and, thus, other tubes is also reduced. For most purposes the vapour phase provides the ideal environment for storing cells.

Liquid phase temperatures ($-196^{\circ}C$) are advisable when the material is to be stored for very long periods, often longer than the life of the scientist! Particularly precious material should also be stored in the liquid phase so that if a leak develops in the freezer and liquid nitrogen evaporates there is still a time delay before the material is lost.

Many $-70^{\circ}C$ freezers and liquid nitrogen containers are fitted with computerised alarms so that engineers can be called out automatically at any time should the temperature begin to rise to dangerous levels.

ΠΠ What safety precautions should be taken to protect both the operator and the stored material when using liquid nitrogen?

The main dangers come from explosions of ampoules or cryotubes. For this reason it is essential that the operator wears appropriate protective clothing (protective goggles, face mask, gloves). Injudicious handling of cold utensils can also cause burns so it is important to use handling tongues. Exploding ampoules also means that the preserved cell culture may be lost. Thus it is imperative to reduce the risk of such losses. One way is to transfer the ampoules to the gaseous phase a few days prior to use. In this way, any liquid nitrogen that has entered the ampoules may evaporate slowly and leak out of the vessels.

ampoules

special
cryotubes

The cells are stored, ideally at a concentration of around 10^6-10^8 cells per ml in tubes designed specifically for very low temperatures. These cryotubes are made of resilient polypropylene and have a rubber 'O' ring in the screw caps to prevent leakages which could lead to explosions upon thawing.

Alternatively, glass ampoules may be used. These, however, need heat-sealing apparatus and considerable skill to ensure that they are sealed properly. They are also quite difficult to open whilst ensuring asceptic conditions are maintained. It is not surprising, therefore, that the screw-topped plastic cryotubes are preferred by most workers.

5.4.2 Preparing cells for freezing

As we have seen cells for freezing are prepared by trypsinisation to detach adherent cells from flask surfaces. Cells in suspension are centrifuged at around 100xg for 5-10 minutes and the pellet is resuspended in a small volume (1-2 ml) of storage medium.

cryoprotectants
glycerol or
DMSO

The composition of the medium used to suspend cells is rather specialised. It contains a cryoprotectant (usually glycerol or dimethyl sulphoxide (DMSO)) and a high protein concentration (usually extra serum). The formulation of this storage medium has evolved through many trial and error experiments.

The usual concentrations of additives are:

- cryoprotectants 7-10% (v/v);

- serum 20% v/v.

use of fresh DMSO

If DMSO is used, care must be taken to avoid direct contact as it is a powerful solvent that can penetrate skin. It is also readily oxidised and the products of oxidation are toxic to cells. Thus fresh DMSO should be used and ideally it should be stored at -20°C.

5.4.3 Freezing is stressful

cells become dehydrated

osmotic effects

When viable cells and tissues are frozen a number of stresses are set up within the intracellular environment. At around 0°C ice crystals begin to form and these crystals grow as the temperature drops. Because the crystals are made of water alone, the result is that the constituents of the cell which are normally dissolved in water become more and more concentrated as the water freezes. This process of dehydration creates osmotic pressures within the cell which are thought to be the major cause of cell damage and death.

importance of controlled temperature drop

The formation of ice crystals is only a problem, however, when the drop in temperature is rapid. A great deal of the damage can be prevented if the rate of cooling from 0°C to -50°C is slow and controlled. A fall of about 1°C per minute is optimal. Once a temperature of -50°C is reached cells must then be cooled rapidly to the final holding temperature.

Rapid chilling results in thermal shock and leads to cell death or injury. Some mechanical injuries to cell membranes during freezing also occur due to ice crystals forming in the extracellular matrix. This mechanism is not thought to be the most common cause of cell death and can be avoided by using a cryoprotective such as DMSO at 7-10% final concentration in the storage medium.

Let us examine this process of freezing in a little more detail.

5.4.4 Slow freezing or controlled freezing is vital

The cells suspended in a small volume (0.5-1.0 ml) of storage medium should be carefully labelled identifying the cell type, approximate cell density and date of storage. There is usually little more room to put any other information on the actual tube but detailed records should be kept on index cards or computer databases.

Sophisticated devices are available to control the freezing of cells at a rate of 1°C per minute. Programmable freezers are useful in laboratories whose main purpose is cell storage but are generally too expensive for the smaller laboratory. A crude but effective alternative is to place the tubes to be frozen in an upright position in a polystyrene box which can then be placed at -70°C overnight or until the cells have frozen to below -50°C. They can then be transferred immediately to liquid nitrogen for rapid cooling to -120°C or -196°C.

5.4.5 Keeping records and labelling

It is vitally important to keep proper records of the preserved cell lines. Remember that these cultures may be kept for a considerable length of time. In many cases, the preserved cultures may be kept long after the original depositor has left.

∏ See if you can make a list of the information that needs to be retained.

The information that needs to be kept is quite extensive. Here we will outline the main features. We have put this information under a number of headings.

- features of the cell line: origin, cell type, cell line number, tests carried out, for example for viruses, mycoplasmas etc;

- date of preservation/storage;

- person responsible for deposition;

- location in the store;

- cell density (approximate);

- number of ampoules/tubes stored;

- number and date of removal of ampoules, details of where the culture has been used - results of viability checks;

- other special features (for example if cited in a patented process).

This type of information is kept on index cards or, increasingly, is stored electronically. You may well imagine that for cell banks (see below) that preserve many thousands of cell lines, keeping records is a major and integral part of cell line preservation.

∏ Explain why the removal of ampoules must be monitored.

replenishment
of stocks

This must be done to ensure that the last ampoules are not removed and the culture lost. In other words, it is essential that when the number of ampoules is reduced to a particular critical level, the stock is replenished by repeating the cultivation-preservation cycle.

The number of ampoules produced and the regularity of re-furbishing these stocks is, of course, governed by the anticipated and actual use of the stocks.

It is also important to know who has used stock cultures. If a subsequent problem arises (for example if the culture is found to be erroneously labelled or contaminated with a virus) the recipient can be traced and alerted.

∏ Why is an approximate cell density recorded?

Without this value, it is difficult to ascertain whether or not the cell line is maintaining viability during storage. It may also be needed for the initial inoculation of culture medium when the samples are recovered from storage.

5.5 Thawing frozen cells

rapid thawing
is best

Because of the problems of genetic instability and contamination mentioned earlier, it is necessary to replace the working stocks from time to time. Cells retrieved from storage must be thawed rapidly to ensure maximum survival. Where possible the ampoule should be plunged into a beaker of water at 37°C until thawed. A face mask or visor and gloves should always be worn to protect skin and eyes from the liquid nitrogen and also for protection in case the ampoule explodes as the liquid nitrogen expands.

As soon as the cells have thawed completely (around 30 seconds) the cell suspension can be transferred dropwise into a centrifuge tube or container containing about 20 ml of pre-warmed growth medium supplemented with 10% foetal calf serum.

∏ See if you can give a reason why cells are added dropwise to the medium?

The main reason is that this approach rapidly dilutes the cryoprotectant. In many cases, the rate of removal of the cryoprotectant is critical to the survival of the cells. Some cells are particularly sensitive to cryoprotectants and it may be necessary to remove the cryoprotectant by immediate centrifugation.

The washed cells can then be transferred to a small $25cm^2$ flask containing fresh growth medium to recover for a few days.

importance of
cell density

The cell density used during resusication is important. If too low a density is used, the cells remain in a lag phase for a very long time. Usually a density of $2\text{-}5 \times 10^4$ cells ml^{-1} is recommended for anchorage-dependent cells and $3\text{-}6 \times 10^5$ cells ml^{-1} is recommended for anchorage-independent cells.

SAQ 5.3

An ampoule containing a cryopreserved cell line has been recovered from the cryostat and warmed to 37°C. After opening, the contents of the ampoule have been dropped into 5 ml of pre-warmed medium. A small sample of this suspension has been stained with trypan blue and examined in a Neubauer counting chamber under a microscope. From this sample, it was estimated that the original suspension contained 8×10^7 cells of which only about 10% were viable. The cells line is normally cultivated in suspension.

1) What volume of medium should be added to the cell suspension?

2) Would you be satisfied with the recovery rate of this cell line? Justify your answer.

5.6 Quantitation of cell viability

It is useful to quantitate the rate of recovery following freezing for two important reasons:

- quantitative assessment of cell viability will give some idea of how soon the cells will recover sufficiently to be used or passaged;

- a low number of viable cells may indicate problems with the storage protocol or temperature failure during storage (see SAQ 5.3).

Here we will examine some of the main techniques for evaluating cell viability.

5.6.1 The dye exclusion test

trypan blue
erthrocin B

This test is crude but gives a rough estimate of cell viability very quickly - so valuable time and reagents are not wasted trying to maintain cells which are already dead. As we have seen in this technique, an isotonic solution of a dye such as trypan blue (0.025%) or erythrocin B is added to a small aliquot of the cell suspension. Cells which have retained membrane integrity will not have taken up dye, whereas those whose membranes have been damaged during the freezing and thawing will have had the dye diffused through into the cytoplasm. Cells which do not take up the dye can be counted using a haemocytometer.

5.6.2 The clone-forming efficiency test

This is a more accurate test and it can be used for adherent cells. An aliquot of the reconstituted cells is diluted to give between 10^2 and 10^4 cells ml^{-1}. Serial tenfold dilutions of this cell suspension are made in growth medium and each dilution is inoculated into small 25 cm^2 flasks. The flasks are incubated at 37°C for 12-14 days after which the cells are fixed with formaldehyde. The fixed cells can then be stained with a simple dye (eg 1% toluidine blue) and examined under the microscope.

The clone-forming efficiency is calculated as:

$$\frac{\text{no of clones formed}}{\text{no of cells inoculated}} \times 100\%$$

∏ What are the advantages and disadvantages of these two methods?

The clone-forming efficiency test gives a measure of the number of cells which actually grow. In contrast, the dye exclusion test merely measures the number of cells which do not take up dye. It is by no means certain that all such cells are capably of growing and dividing.

The advantages of the dye exclusion test are that it is simple, requires little specialist equipment and gives a rapid result. The clone-forming efficiency test, however, requires substantial media preparation and the results are only obtained after 12-14 days.

5.7 Cell banks

There are a large number of culture collections which maintain large stocks of animal cells. These collections fulfil a number of functions. Most importantly they provide resource centres from which cell lines may be obtained. This is a two way process. Scientists can submit the cell line they have obtained from their own research. Alternatively they can obtain cultures for research or teaching purposes.

services of the major culture collections

The culture collections also provide a variety of other services. For example they will characterise cell lines and, in some instances, provide training resources for the maintenance and characterisation of cell lines. They also act as depositions for patented cell lines. We will not go into details of the patenting process here, except to say that patenting procedures demand that relevant cell lines are submitted to a recognised cell culture organisation. (Details of patenting involving organisms and cell lines are dealt with in the BIOTOL text 'A Compendium of Good Practices in Biotechnology')

Many culture collections also provide consultative services, offering advice on a wide range of cell cultivation issues. We have provided a list of some of the major collections in Table 5.1. These are listed in alphabetical order according to their geographical location.

Australia	Commonwealth Serum Laboratories 44 Poplar Road, Parkville, Victoria 3052, Australia
Bulgaria	The National Bank for Industrial Micro-organisms and Cell Culture (NBIMCC), Blvd Lenin 125 BL2, V Floor, Sofia, Bulgaria
Canada	The Repository for Human Cell Strains and Cell Repository for Neuromuscular Disease The McGill University, Mountreal Children's Hospital Research Insititute, 2300 rue Tupper Street, Montreal, H3H 1P3 Canada
European: Collection of Biochemical Research	
	Department of Cell Biology and Genetics, Erasmus University, PO Box 1783, NL-3000 DR Rotterham, The Netherlands
	Institute Nazional par la Recerca sul Cancro, Viale Benedette XV, 10, 1-16132 Genova, Italy
	Institut fur Immunogenetik, Universitat Sklinkum Essen, Virchowstrasse 171, D-4300 Essen 1, Germany
	European Human Cell Bank, ECACC PHLS Centre for Applied Microbiology and Research, Porten Down, Salisbury, SP4 0JG, UK
France	Collection Nationale de Cultures de Micro-organisms, Institut Pasteur, 25 rue du Dr Roux, F-75724 Paris Cedex 15, France
Germany	Tumor Bank, Deutsche Krebs forschungszetrum, Institut fur Experimentelle Pathologie (dkfz), Im Neueriheimer Feld 280, Postfach 101949, D-6900 Heidelberg 1, Germany
Hungary	National Collection of Agricultural and Industrial Micro-organisms (NCAIM), Department of Microbiology, University of Horticulture, Somloi ut 14-16, H-118 Budapest, Hungary
Italy	Centro Substrati Cellulari, Institute Zooprofilattico Spermentale della Lombardia e dell Emilia Via A Biandri 7, 1-25100 Brescia, Italy
Japan	Institute for Fermentation, Osaka, 17-85 Jusan-honmadu 2-chome, Judagawa-ku, Osaka 532, Japan
	Patent Micro-organsim Depository Fermentation Research Institute Agency of Industrial Science and Technology, 1-1-3 Higaslu, Yatabe, Tsukuba Science City 305, Japan
	Patent Micro-organsim Depository Fermentation Research Institute Agency of Industrial Science and Technology, 1-1-3 Higaslu, Yatabe, Tsukuba Science City 305, Japan
	(Note that in Japan there are cell culture collections held at the Universities of Tokyo, Yokohama, Kyoto, Okayama)
United Kingdom	European Collection of Animal Cell Cultures (ECACC), PHLS Centre for Applied Microbiology and Research, Porton Down, Salisbury SP4 0JG, UK
United States of America	American Type Culture Collection of Cell Lines, 12301 Parklawn Drive, Rockville, Maryland 20852, USA

Table 5.1 Major collections of cell lines.

restriction on
deposition and
acquisition

You should realise that a number of conditions are attached to the deposition and acquisition of cell lines. Although the major aim of culture collections is to make cultures available to the scientific community, the conditions of release are governed by concerns of safety and the commercial value of the cell lines. Patented organisms are not generally released. Likewise the release of cells which present some clinical danger are usually restricted to those who can demonstrate that they have appropriate containment faculties and expertise.

Almost all culture collection services have to offset some of their operational costs by charging for the supply of cultures and they all operate within the legal framework for transporting cultures. This includes import and export requirements. Also of concern is the need to ensure that cultures remain viable during shipment.

quality
assurance

In all culture collections, a key question is quality assurance to ensure that cultures are not contaminated and that cultures received from external sources are exactly as the depositor believe them to be. Equally, recipients of a culture need to be assured that the culture they receive is properly accredited.

Cell culture collections use a variety of procedures to ensure the quality of supplied cultures, but a generally accepted procedure is as described in Figure 5.4.

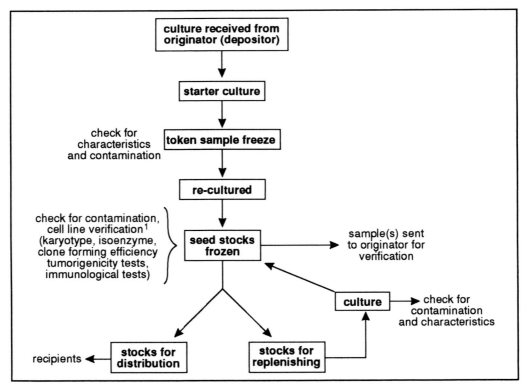

Figure 5.4 Schematic representation of the procedure used to verify and maintain cell lines in cell culture collections. 1) Many of these verification tests are described in Chapter 4. Note that the characterisation tests used are to some extent dependent upon the cell line.

SAQ 5.4	Suggest reasons why you should deposit a cell line you have isolated and characterised with a cell culture collection. Give reasons for not submitting a cell line immediately you have isolated it.

SAQ 5.5	1) A cell line you are interested in is available from a number of culture collections including:

Commonwealth Serum Laboratories, Victoria, Australia;

Department of Cell Biology and Genetics, Erasmus University, Rotterdam, The Netherlands;

European Human Cell Bank, Porton Down, Salisbury, UK;

American Type Culture Collections of Cell Line, Maryland, USA.

List the factors which would influence your choice of the culture collection you would apply to for a sample of the culture.

2) Describe what you would do on receipt of the culture.

Summary and objectives

In this chapter we have explained why an effective method of storage of cell lines is required and the techniques used to preserve cell lines are described. The preservation of cell lines depends upon selecting suitable cells, using properly formulated media and freezing the cells under controlled conditions. Critical to the success of these processes is the use of appropriate freezing and thawing regimes. Usually cells are preserved at liquid nitrogen temperatures (-196°C) or in nitrogen vapour at -120°C. We also examined the techniques of determining viability.

In the final part of the chapter, we described the major culture collections which act as valuable resources of cell lines and advice.

Now that you have completed this chapter, you should be able to:

- describe the problems of genetic instability in cell lines;

- describe the processes which occur in the cells and their environment as the temperature is reduced from 0°C to -50°C;

- discuss the merits and demerits of preservation of cells at -120°C and -196°C;

- outline the safety precautions necessary for handling material stored in liquid nitrogen;

- list the criteria used in selection of suitable cells for cryopreservation;

- explain the techniques used to evaluate cell viability (the dye exclusion test and the clone forming efficiency test);

- describe the roles of the major cell culture collections.

Oncogenes and cell transformation

6.1 Introduction   116

6.2 Cancer and cell transformation   116

6.3 Cell transformation   118

6.4 Cancer is caused by changes in the genome   120

6.5 The molecular basis of cancer   121

6.6 Oncogenes   122

6.7 Tumour suppressor genes   127

6.8 Multi-stage carcinogenesis   129

6.9 The importance of cancer research to biotechnology   129

Summary and objectives   131

Oncogenes and cell transformation

6.1 Introduction

tumours
(neoplasms)
and cancer

During animal development, cell division and cell differentiation are carefully regulated, but occasionally the controls that regulate cell multiplication break down and cells begin to grow and divide although the body has no need for them. These cells have the ability to divide without responding to regulation, thus producing groups of cells which can expand indefinitely. A tumour is such a mass of cells which have escaped from the normal mechanisms controlling cell growth. Tumours are also called neoplasms, which means literally 'new growth', but has come to mean abnormal new growth of tissue. Likewise tumour cells are sometimes called neoplastic cells, or hyperplastic cells. The term **cancer** is used to describe a large group of diseases which arise because of changes in growth and behaviour of cells.

reasons for
studying
neoplastic cells

The study of neoplastic cells and the cancers they produce is of fundamental importance. Cancers are amongst the most feared medical conditions and studies on how they arise may enable the development of strategies for preventing and curing these conditions. As we will learn later, the study of neoplasia has, to a large extent, depended upon using cells cultivated *in vitro*. This alone justifies inclusion of a chapter on cancer in this text. There is however a further justification. Neoplastic cells are cells that have escaped from normal regulatory control. The study of these cells may, therefore, provide insights into how cell growth and division are regulated in normal cells. This has potential practical implications in the use of cultivated animal cells to make biotechnological products such as specific mammalian proteins. The enhanced growth rates and ability to grow in relatively simple media, both features of neoplastic cells, are, of course, desirable features for the use of animal cells cultured *in vitro* for commercial purposes.

In this chapter, we will first examine the biology of cancer and the mechanisms by which neoplasms arise. Towards the end of the chapter, we will comment on the potential applications of this knowledge to the cultivation of animal cells *in vitro*.

It is difficult to study cancer without some knowledge of normal growth. We anticipate that readers will have a basic understanding of the normal developmental controls which take place during embryogenesis and adult development. If, however, you are not confident in this area, we recommend the BIOTOL text 'Infrastructure and activities of cells'.

6.2 Cancer and cell transformation

clinical
consequence
of cancer

metastasis

The unpleasant symptoms of cancer result from the unlimited growth of the neoplastic cells, which results in an increase in the bulk of tissue, impinging on the structure and function of adjacent tissue. In the later stages of cancer, the cells of the tumour can penetrate and invade adjacent tissues, spreading from the primary site to distant sites in the body. It is this ability to spread (metastasize) that characterises malignant tumours. As a cancer progresses, there are changes in the histology of the tumour which include the tendency of the cells of the tumour to lose their differentiated properties. From a prognostic stand-point, patients with poorly differentiated tumours generally have a lower survival rate than those with well differentiated tumours.

clinical
prognosis of
cancer

Despite tremendous efforts, there has been little improvement in the overall survival of patients with cancer over the last 30 years. Whilst changes in the management of lymphomas and leukaemias have produced a dramatic increase in survival, these contribute little to the overall picture. Around one person in three will develop cancer at some time during his or her life. It is likely that the overall incidence of cancer will continue to rise as the average age of the population increases. Investigations into the causes, prevention and treatment of cancer is, therefore, a major component of scientific research.

There are many ways of studying cancer ranging from clinical trials involving patients with cancer to the in-depth analysis of the very early events which take place in the cells which go on to form tumours. The use of tissue culture has been indispensable in the latter course of study and it is this approach which will be presented here.

6.2.1 Normal growth and transformation

Normal growth and cancer are two faces of the same coin, and much of cancer research involves the study of controls on normal cell growth, in the hope that we will eventually understand how these normal controls have been disrupted in the cancer patient. In the normal cell, growth is controlled by a variety of positive and negative control pathways and these are finely balanced to induce growth only when required. An example of a positive control pathway which stimulates growth is the well controlled cascade of events that occurs when a mitogen binds to its receptor and induces cell proliferation. Negative control pathways exist to ensure that each step along this pathway is switched off when it is no longer needed. However, in this complicated scheme of stimulatory and inhibitory pathways there is much scope for errors to occur, and occasionally they do. The result is that cells lose their developmental controls and start to proliferate when they should remain quiescent. In other words, cancer can be viewed as a deregulation of the fine balance between positive and negative growth control mechanisms.

positive control
through
mitogens

This loss of growth control is associated with a number of measurable changes in the phenotype of cells and when a cell acquires these phenotypic properties it is said to be transformed. We will examine the properties of transformed cells in detail a little later.

6.2.2 'Normal' and non-transformed cells

Research into the molecular basis of cancer depends heavily on cultured cells, although clearly this work has to be complemented by *in vivo* experiments. Cultured cells have the advantage that the environment of the cells can be manipulated by the investigator, the type of target cell can be well defined, and changes in cells following treatment with carcinogens can be investigated. Furthermore these cells can be manipulated genetically.

∏ The word 'normal' is sometimes used to describe cells lines obtained from normal tissue. Why is this inappropriate?

In Chapter 1, we stated that most diploid cells have a limited life span, and only rodents routinely give rise to continuous cell lines from normal tissue. Although these cell lines maintain many of the characteristics of the cell type *in vivo*, they cannot really be described as normal for a number of reasons. These cells tend to divide faster than cells *in vivo*, they have a slightly reduced cell size, and more importantly, they are genetically different from the original cells, since the culture is aneuploid. However, if these cells are maintained at low cell density, they retain many of the characteristics of normal cells. It is useful at this stage to make the distinction between 'normal cells' and 'non-transformed cells'. The designation of cells as normal implies that growth control is normal, that the cells have undergone no genetic changes, and that the cells have not been transformed. By these standards, immortalised rodent cell lines are not normal. They are, however, very useful for the study of cellular transformation since many of these cell lines are non-transformed. Treatment of such adherent cells with various agents (viruses, chemicals, radiation) can dramatically change the growth properties of cells in culture: the cells become transformed.

In the next section, we will describe in more detail exactly what is meant by the term 'transformed' and we will then look at some of the recent advances which have been made using animal cell culture as a model for cancer induction *in vivo*.

6.3 Cell transformation

Transformation is defined as the acquisition by cells of certain phenotypic properties associated with cancer. A number of assays have been developed in order to establish whether cells are transformed and they have generally proved reliable if used in the correct context. Transformed adherent cells be identified by changes in the control of their growth and behaviour, and some of these changes are described in more detail below.

6.3.1 Phenotypic properties of transformed cells

The changes to cells when they become transformed can be treated under five headings:

* ability to form foci in *in vitro* culture;

* changes in morphology;

* changes in growth factor requirements;

* release from anchorage-dependence for cell division;

* ability to form tumours *in vivo*.

These changes provide the basis for determining whether or not cells have been transformed. We will discuss each in turn.

Focus formation

The most obvious and medically significant property of cancer cells is that they proliferate uncontrollably. When normal cells are grown in culture, they form a monolayer and when this monolayer reaches confluence the cells stop dividing through a process called contact inhibition. Transformed cells loose this property, and produce

normal vs non-transformed cells

contact inhibition

disorganised clusters of cells which grow on top of each other. These are called 'foci' (singular, focus), and these are readily recognisable when seen against a background of normal cells growing as a monolayer. This assay is often referred to as the focus forming assay (see Figure 6.1a).

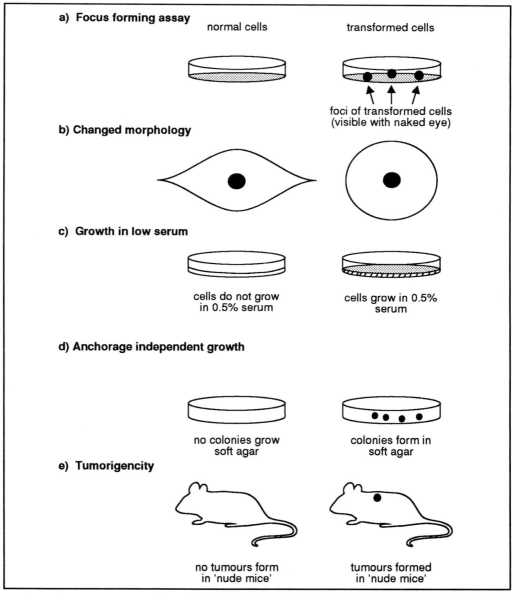

Figure 6.1 Diagrammatic illustration of the techniques used to demonstrate the presence of transformed cells in a culture (see text for discussion).

Morphology

The morphology of cells changes upon transformation. Usually transformed cells tend to round up, probably due to the disruption and reorganisation of cytoskeletal structures (see Figure 6.1b).

Growth factor requirement

Transformed cells proliferate under conditions in which normal cells do not. When such cells are cultivated *in vitro* they will grow in media containing low serum concentrations whereas non-transformed cells will not do so (see Figure 6.1c). This reduced requirement for exogenous mitogenic stimuli is now known to result from abnormal expression of growth factors or growth factor receptors or by subversion of the signal transduction pathways involved in normal growth control.

Transformed cells become anchorage-independent

You may recall from Chapter 1 that most cells (with the exception of haemopoietic cells) grow as an adherent monolayer. Adherent cells are said to be anchorage-dependent which means that they need to attach to a substrate in order to proliferate. When cells are transformed, they acquire anchorage-independence, the ability to grow without attachment to a substrate. In practice, when cells are suspended in soft agar, growth of colonies from single cells in such a medium indicates that the cells have become transformed (see Figure 6.1d).

Tumorigenicity can be demonstrated in 'nude' mice

nude (athymic) mice

The ultimate criterion for cell transformation is tumorigenicity: the ability to generate tumours. This can only be demonstrated *in vivo*, and is measured experimentally using a special kind of mouse called the 'nude' mouse (also called the athymic mouse). The nude mouse is a developmental mutant which has lost its ability to produce a thymus gland and therefore has a defective immune defence system. It can therefore serve as a host for the growth of tumorigenic cells originating from any species, since these cells will not be recognised as foreign by the animal's immune response. The rate of tumour formation and the number of tumours formed when cells are injected under the skin of the immune deficient mouse gives an indication of tumorigenicity.

SAQ 6.1

Name three assays suitable for analysing the presence of transformed blood cells in culture.

6.4 Cancer is caused by changes in the genome

The phenotypic differences between non-transformed and transformed cells covered above provide only a descriptive account of transformation without explaining how or why transformation takes place. One clue as to the causes of transformation is the fact that these phenotypic changes are a reflection of genetic changes. The genetic material of a cell is somehow changed and cell populations with these genetic changes are generated by growth and division.

The idea that cancer may be essentially a genetic disease dates back to 1914, when Boveri suggested that chromosomal abnormalities might convert a normal cell into a cancer cell. However, it has largely been over the last two decades that this idea has been substantiated and the exact nature of these genetic changes has been characterised in more detail.

This relatively recent insight regarding the cause of cancer was dependent upon advances in a number of different branches of science. In the mid 1970s the development of recombinant DNA technology led to a radical new approach to biological research. Recombinant DNA techniques permit the formation of new combinations of genes by allowing us to isolate a DNA fragment, change it in a defined way and introduce it into the same organism or a different organism. Thus it became theoretically possible to isolate genes responsible for transformation and study them in detail. The study of viruses actually led to the discovery of such genes. The first gene identified which could transform a normal cell, causing the phenotypic changes associated with cancer, was isolated from a virus. Subsequently such genes, called onocogenes, were isolated from other sources including human tumours. Further study

discovery of oncogenes and proto-oncogenes

indicated that oncogenes were mutated versions of normal cellular genes which play a role in normal growth control. The normal genes were therefore called proto-oncogenes. In summary, proto-oncogenes play a role in normal growth control but certain mutations of such genes result in subversion of this growth control leading to cancer.

It is important to remember that cancer reflects changes in the genome (ie it may be regarded as a genetic disease). It is not necessarily inherited. In fact, inherited forms of cancer are rare, although many cancers do show familial clustering.

∏ What is the difference between an inherited disease and a genetic disease?

A genetic disease involves changes in the genetic material of the cell, in other words, in the cell's DNA. When such a change (a mutation) occurs in the somatic cells of the body, it will not be transmitted to subsequent generations. Only mutations in the germ-line cells will result in the mutated genes being present in all the cells of the offspring and will be carried to each subsequent generation. In an inherited disease the mutation is present in the germ-line cells and therefore passed on to the next generation. Thus cancer is a genetic disease but not necessarily an inherited disease.

6.5 The molecular basis of cancer

A very important goal in the field of cancer research is the identification of genes that convert a normal cell into a tumorigenic cell and the genes that suppress the phenotypic properties of a tumorigenic cell. Molecular biology has opened the door towards the definition of three major groups of genes involved in neoplasia. These are:

Oncogenes which act dominantly in the sense that they are able to induce at least some of the changes in the cell phenotype associated with the neoplastic state when transferred into cultured cells. An oncogene is defined as a gene that contributes to neoplastic transformation when it is introduced into a non-transformed cell;

Tumour suppressor genes which have to be deleted or functionally incapacitated before a tumour can arise;

Modulators which modify the spread of neoplastic cells in the organism, for example, by influencing metastatic spread, invasiveness, or cellular resistance to immune rejection. Many potential cancer cells are prevented from causing neoplasia by the immune system, which recognises and destroys the transformed cells. The importance of immune surveillance in keeping potential tumours at bay is demonstrated in patients who are immunodeficient. In these patients the eventual cause of death frequently involves tumour growth.

We will examine each of these types of genes in turn but place most emphasis on oncogenes and tumour suppressor genes.

6.6 Oncogenes

We will begin by describing the discovery of oncogenes and the related proto-oncogenes. We will then explain the role of proto-oncogenes in normal cells before describing how proto-oncogenes can be activated to produce oncogenes.

6.6.1 The discovery of oncogenes

v-src from RSV

In 1911, Peyton Rous demonstrated that cell-free filtrates of certain chicken sarcomas (tumours of connective tissue) promote new sarcomas in chickens. Interestingly, Peyton Rous was one of many people who did not believe that genetic mutations played a role in the genesis of cancer, but the virus he discovered eventually led to the proof that genetic changes can cause cancer. The virus which had caused these tumours was later identified as Rous Sarcoma Virus (RSV). RSV is a retrovirus (an RNA virus that replicates itself by copying its RNA into DNA, inserting this DNA into the host cell genome, and then transcribing the DNA). The first oncogene to be discovered, *v-src*, was identified from RSV, and its protein product, a 60 kD phosphoprotein (called $pp60^{v\text{-}src}$) was found to be responsible for the induction of tumours and neoplastic transformations by RSV.

6.6.2 The discovery of proto-oncogenes

cellular *c-src* gene homologous to *v-src*

In 1976, Michael Bishop and Harold Varmus made the remarkable discovery that uninfected chicken cells contain a gene which is homologous to viral *v-src*. This cellular gene was called *c-src*. Moreover, species as divergent as humans and the fruit fly *Drosophila* all contained a similar gene. This raised the question 'what were these potential cancer-inducing genes doing in the normal genome?' We will attempt to answer this.

6.6.3 The role of proto-oncogenes

v-sis homologous to PDGF

In 1983, there was a breakthrough in the understanding of the role of proto-oncogenes. By this time a number of different oncogenes and proto-oncogenes had been identified and one oncogene, called *v-sis* (found in Simian Sarcoma Virus, SSV for short) which had been sequenced, was found by computer searches to be homologous to the gene encoding one of the chains of human platelet-derived growth factor (PDGF). In other words, the gene coding for this PDGF chain was the proto-oncogene corresponding to the *v-sis* oncogene. This suggested that proto-oncogenes are genes encoding proteins which play a role in normal growth regulation. A non-transforming ancestor of SSV had picked up, or hijacked, the normal PDGF gene from the cellular genome by a process known as genetic transduction.

The PDGF gene, now in a different genetic 'environment' was not under the normal regulatory control. Thus, when the PDGF gene-carrying virus infected a host cell, the PDGF gene was expressed and thus any cells with PDGF receptors were stimulated to grow. We have represented this in Figures 6.2 and 6.3.

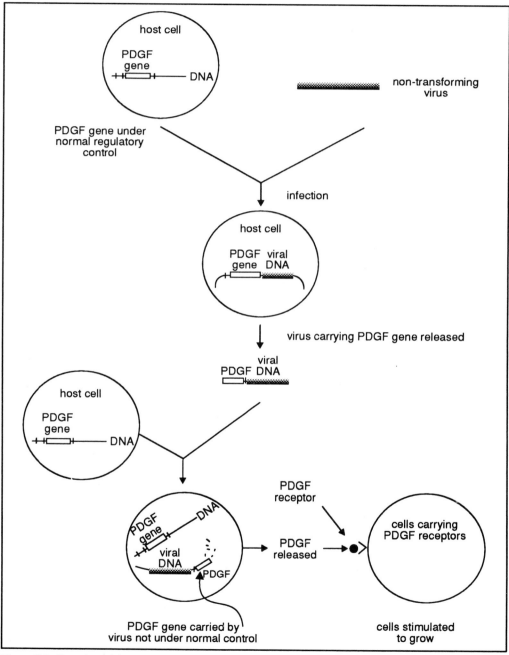

Figure 6.2 Overview of the process by which a non-transforming ancestor of SSV may have picked up the PDGF gene to become a transforming virus (see text for a description).

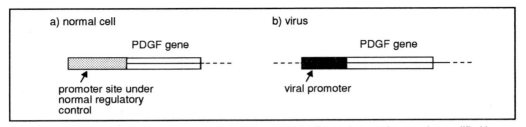

Figure 6.3   A simplified figure to show how the regulation of PDGF gene expression may be modified by being placed in a different genetic environment. a) In a normal cell, the PDGF gene is under normal regulation; it is controlled by regulatory sequences (eg promoters, enhancers) adjacent to the PDGF gene. b) When the gene is incorporated into a viral genome, it may be separated form its normal regulatory sequences. Thus its expression is no longer regulated in the normal way.

v-erbB and
v-fms are
oncogenes

Other viral oncogenes were subsequently found to be related to different genes involved in the normal growth control pathway eg, *v-erbB* (of avian erythroblastosis virus) is related to the epidermal growth factor (EGF) receptor and *v-fms* is related to the receptor for colony-stimulating factor-1 (CSF-1). There are many more examples.

Let us examine whether or not you have understood the principle of transformation by retroviruses which have picked up a proto-oncogene by attempting the following SAQ.

SAQ 6.2

Let us assume that a retrovirus has picked up the gene which codes for the receptor for the epidermal growth factor and that this gene is under the control of a viral promoter. What might be the consequences if this virus infects a thyroid cell?

As we have indicated before, the regulation of growth and division of mammalian cells is complex. We can, however, attempt to represent this regulation as a simple cascade in the following way:

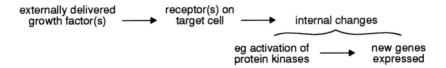

In principle any gene carried by a virus which bypasses or changes any of these steps will lead to changed growth characteristics. For example if the gene carried by the virus produces a growth factor, this will stimulate growth of the target cells of the growth factor. Equally, if the virus carries a gene whose product activates the key protein kinase, this too will lead to changes in growth characteristics. We might, therefore, anticipate that the range of possible genetic mechanisms for bringing about transformation may be quite extensive.

Although much is known, there is still much to be done to establish the mechanism of transformation mediated by many viruses.

This is a very vibrant area of research and you should anticipate that much more will be established over the next few years. This area of research is, however, complicated by a rather difficult nomenclature. The names given to many of the genes which are involved in normal growth control are confusing since many of these genes were discovered by studying systems in which they had gone wrong. For example, the *c-src* gene is involved in normal growth control but owes its name to the sarcomas that the oncogene *v-src* causes in chickens.

6.6.4 Are all oncogenes viral?

In the previous section we established that the oncogenes were discovered in retroviruses and that these genes were derived from host proto-oncogenes. The question we might ask is 'are all cancers caused by viruses?' Perhaps you would like to answer this question before reading on.

chemicals and radiation may induce cancer

The answer is, of course, that not all cancers are caused by viruses. You should be aware, for example, that many chemicals (mutagens/carcinogens) and high energy radiation (for example X-rays) are capable of inducing cancer. In humans, for example, there is little evidence for cancers caused by retroviruses.

oncogenes from tumours shown by transfection

Tumours which are not caused by viruses have also been shown to contain oncogenes by a technique known as transfection (see Chapter 10). Most non-viral oncogenes were identified by transfection of DNA from human cancer cells into mouse 3T3 cells (see Chapter 1). This technique results in transformation of cells, and the gene which caused the transformation can subsequently be isolated and studied in detail. One group of oncogenes, called *ras* genes, has been found in 30% of human cancers. Like the viral oncogenes, these oncogenes were also found to have homologous proto-oncogenes in normal cells only these proto-oncogenes were activated into oncogenes by different mechanisms. We will examine these mechanisms later.

6.6.5 Classification of oncogenes and proto-oncogenes

In order to find out the normal function of proto-oncogenes and to establish how changes in proto-oncogenes lead to neoplasia, the physiological functions of proto-oncogene products have been studied extensively. Most proto-oncogene products have been found to fall into categories of proteins which were known for their involvement in normal growth control. From our simple control cascade described in section 6.6.3 you should realise that some proto-oncogene products resemble known growth factors, others code for growth factor receptors, yet others for intracellular signal transducers, and finally some resemble nuclear transcription factors. This finding exemplifies how important it is to study normal growth as well as the events which occur upon transformation.

6.6.6 Mechanisms of activation of proto-oncogenes giving rise to oncogenes

Proto-oncogenes can be activated by a number of different mechanisms, some of these have already been mentioned. Here they are grouped together into different classes.

Changes which result in an altered protein

Even single point mutation can render a gene oncogenic by changing the structure and function of the protein which it codes for. For example the *ras* proto-oncogene which was mentioned above is often activated into an oncogene by a single point mutation. In this particular example the product of the gene sends signals instructing the cell to divide in an unregulated manner so that the cell divides when it should not.

Π   Where does a point mutation have to occur in order for a proto-oncogene to become an oncogene? [select from a) in the structural gene or b) in the regulatory sequences controlling the expression of the oncogene].

The answer is that in principle it could be either. A point mutation within the structural gene may lead to the production of a protein with changed properties. For example it might bind more, or less, strongly with its receptor. Alternatively if the protein is a receptor for a growth factor, it might be able to activate the next step in the control cascade even in the absence of the growth factor.

We can represent this in the following way:

a) normal cell

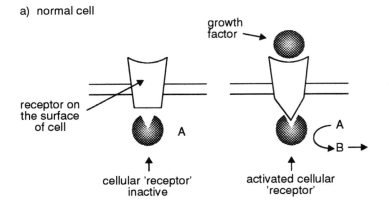

b) cell with modified receptor

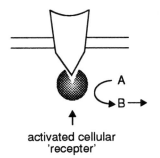

In the normal cell a), events in the cell are regulated by the presence/absence of the external growth factor. In b) the 'modified' receptor activates cellular processes even in the absence of the growth factor.

If the point mutation occurs in the regulatory sequence of a key proto-oncogene, it may be that the expression of this gene is no longer sensitive to normal control. Thus the gene may become over- or under-expressed. This may have quite a profound effect on the growth of the cell. For example, we might get over-expression of a gene coding for a growth factor or a growth factor receptor. It might also lead to the expression of, for example, a growth factor receptor in the wrong cells.

use of PCR to
analyse small
changes in
genes

The recently developed PCR (polymerase chain reaction) technique is ideal for the study of small changes in the DNA of human tumours since it allows the expansion and analysis of small sections of the genome. (This technique is described in detail in the BIOTOL text 'Techniques for Engineering Genes').

Insertional mutagenesis

disruption of
regulatory
sequences by
insertion

Avian leukosis virus (ALV) does not contain any oncogenes, but when it integrates into the host chromosome it often inserts near to a proto-oncogene called *c-myc*. This results in over-expression of the *c-myc* gene product in the nucleus which results in uncontrolled growth. Effectively what happens is that the regulatory sequences of the *c-myc* gene are disrupted.

Chromosomal translocation

changes in
gene
regulation
accompanying
gene
translocation

In 90% of patients with Burkitt's lymphoma, the proto-oncogene *c-myc* is activated by translocation from its normal position at the end of chromosome 8 to the end of chromosome 14, adjacent to certain immunoglobulin genes. This brings *c-myc* under the transcriptional control of the immunoglobulin regulatory sequences, which results in over-expression of the *c-myc* protein in lymphoid cells. In addition the immunoglobulin genes undergo germline rearrangement in order to generate antibody diversity and the location of *c-myc* in this region of the genome results in a high somatic mutation rate of the oncogene.

DNA amplification

cancers
involving
oncogene
amplification
can have poor
prognosis

Amplification of large parts of chromosomes can be observed cytogenetically whereas amplification of specific genes can be observed by Southern blotting. Gene amplification is frequently accompanied by over-expression of the gene product. For some types of cancer there is a correlation between poor prognosis for the patient and amplification of specific oncogenes. The finding that many copies leads to poor prognosis is in itself a useful discovery since it helps to decide the treatment for patients with cancer, and eliminates unnecessary treatment for those with good prognosis.

Mutations in the regulatory sequences of genes can also result in over-expression of the gene product, with similar results as gene amplification (see protein changes above).

It should be self evident that any agent that might bring about changes in DNA sequences is potentially able to convert proto-oncogenes into oncogenes. The mechanisms described above explain why chemicals which cause nucleotides to be inserted in the wrong place in DNA are potentially carcinogenic. The excision from and replacement of nucleotides in DNA damaged by high energy radiation is also error prone. The association of cancers with nuclear radiation can also be accounted for by the mechanisms described above.

We now turn our attention to another set of genes involved in neoplastic transformation, the tumour suppressor genes.

6.7 Tumour suppressor genes

A tacit assumption in the early studies of cancer was the dominant nature of activated oncogenes. However, earlier studies had clearly shown that inactivation of tumour suppressor genes could also play a role in transformation. It is harder to identify tumour suppressor genes than oncogenes because of the very nature of their function.

Π Explain why it might be harder to identify a tumour suppressor gene than an oncogene.

An oncogene contributes to neoplastic transformation when it is introduced into a non-neoplastic cell. It is easy to spot the change when one cell starts to divide in an uncontrolled manner, since visible clumps of cells form. Imagine what would happen if one cell acquired a tumour suppressor gene. The ability to cause reversal of uncontrolled growth would be much harder to observe.

Our understanding of tumour suppressor genes, like our understanding of oncogenes, is heavily dependent on the use of cell culture models. Some of the important experiments are described below.

6.7.1 Somatic cell hybrid studies reveal tumour suppressor genes

tumour suppressor genes shown by somatic cell hybridisations

In 1969, Harris and Klein demonstrated that when malignant cells are fused with normal cells the resulting hybrid cells were non-tumorigenic. They reasoned that this tumour suppression indicated that genes from a normal cell might replace a defective function in the cancer cell and render it 'normal'. Such genetic elements have been called tumour suppressor genes. These findings were controversial for a while, because other researchers reported the opposite: that the hybrids were tumorigenic. It is now understood that their findings were due to the chromosomal instability of the hybrid cells. Chromosomes are rapidly lost from these hybrid cells, and the tumorigenic segregants have a positive growth advantage over other cells and are therefore selected for. However, when these technical problems were overcome, it was clear that many types of tumour do arise through loss of critical growth regulating genes from the genome.

6.7.2 Clinical studies

retinoblastoma Rb-1 gene is a tumour suppressor gene

Further evidence for tumour suppressor genes came from clinical studies of inherited cancer susceptibilities. The best studied example of inherited cancer is retinoblastoma, where a loss of gene function appears to play a role. Children who inherit one defective copy of the Rb-1 gene develop, on average, 3 retinoblastoma tumours each. In 1971 Knudson produced a theory which he called the 'two-hit' theory in which he proposed that all types of retinoblastoma involve 2 separate mutations. In sporadic retinoblastoma both mutations occur somatically in the same retinal precursor cell, whereas in heritable retinoblastoma, one of the mutations is germinal and the second is somatic. So in inherited retinoblastoma, the first event is the acquisition of the Rb-1 defect before birth. The second event is the deletion or mutation of the remaining Rb-1 gene.

6.7.3 Molecular genetics provides the ultimate proof of the existence of tumour suppressor genes

The ultimate proof that deletion of a tumour suppressor gene plays a role in tumour progression must involve the identification of the genes involved. Introduction of the gene into a cancer cell that lacks a functional copy should restore the phenotype to normal. In practice the phenotype of cells derived from tumours can rarely be expected to revert to a completely normal phenotype, since a large number of changes may have occurred in the genome. Only a reduction in the malignant phenotype can be expected by correcting one defect.

Since the development of recombinant DNA techniques, molecular geneticist have focused their attention on malignancies with defined chromosomal abnormalities. Molecular cloning of tumour suppressor genes is very difficult and therefore the list of tumour suppressor genes which have been identified is inevitably much shorter than the list of oncogenes.

As mentioned already, one tumour suppressor gene which has been identified is the *Rb*-1 gene which is deleted in retinoblastoma. The infection of *Rb*-1 deficient retinoblastoma cells in culture with a retroviral construct containing a full length *Rb*-1 cDNA provides the final proof that *Rb*-1 is really a tumour suppressor gene. The outcome shows that expression of the *Rb*-1 gene product results in growth suppression *in vitro*.

6.8 Multi-stage carcinogenesis

There is now considerable evidence that the activation of a cellular proto-oncogene to an oncogene and the deletion of tumour suppressor genes can both be causal events in carcinogenesis. However, single oncogenes under most circumstances are insufficient to cause full transformation. Instead transformation apparently requires a cell to have undergone multiple genetic lesions, which include both activation of oncogenes and inactivation of tumour suppressor genes. It has been thought for many years that the development of cancers takes place in many stages. The current theory of carcinogenesis is that it is a multi-stage process. The activation of oncogenes and inactivation of tumour suppressor genes can cooperate, resulting in fully transformed cells.

SAQ 6.3

1) UV light causes adjacent thymidine residues on DNA strands to form dimers. The cell's DNA repair mechanisms excise these dimers and replace them. Sometimes incorrect nucleotides are inserted. Use this process to explain why individuals exposed to high intensity sunlight over long periods often develop skin cancers.

2) Try to describe a mechanism by which exposure to UV radiation might give rise to growth abnormalities in organs other than skin.

6.9 The importance of cancer research to biotechnology

It should be clear to you that the techniques of contemporary biotechnology have made important contributions to our understanding of the mechanisms of cancer induction. The ability to identify, clone and sequence genes has, for example, enabled us to identify and compare proto-oncogenes and oncogenes. This, however, is not a one way process. Cancer research has provided a focus for the application of biotechnological techniques and has provided an important impetus to the development of biotechnological expertise. In addition, discoveries within the context of cancer research have contributed significantly to our understanding of how growth and gene expression is regulated in normal cells. This increased understanding has potential applications leading to improved cultivation of animal cells *in vitro*. The *in vitro* cultivation of animal cells for use in toxicity and drug efficiency testing and in the production of vaccines, antibodies, growth factors, hormones etc are important commercial processes.

Improvements in the cultivation of animal cells, derived from the discoveries made in cancer research, could, therefore, have important consequences to the commercial success of these operations. (We will discuss these commercial operations in Chapter 8).

The discoveries made in cancer research have another important bearing on the biotechnological industries by providing new targets for the application of biotechnology. By understanding how cancers develop, it becomes possible to identify new approaches to treating these conditions. The production of targeted drug delivery systems to treat cancers depends upon harnessing biotechnological processes. Similarly cancer research may enable us to identify potential 'gene therapy' approaches to solving problems associated with some cancers. For example, we described the high incidence of retinoblastoma in individuals who carry a defective copy of the tumour suppressor gene Rb-1. This raises the possibility of using a genetic approach to reducing the incidence/consequences of this defect. Such an approach would, of course, be dependent upon the application of biotechnological techniques.

Another important area of biotechnology is in the production of diagnostics. The identification of, for example, transformed cells is important in cancer diagnosis and treatment. Again, the biotechnology industry has the potential to generate appropriate diagnostic reagents. In the same context, diagnostic reagents can be used for embryo screening to identify embryos carrying genes which make them susceptible to generating particular cancers. Biotechnology has the potential to generate, for example specific gene probes which enables us to identify if particular genes are present in an embryo/foetus. For example, if a probe was produced which enabled us to identify a defective tumour suppressor gene Rb-1 in an embryo at an early stage, it might facilitate treatment. There are, of course, many ethical issues associated with such courses of action which need to be resolved. We do not propose to discuss these at this stage since our objective is to illustrate how biotechnology has contributed to cancer research and how cancer research may contribute to the development of biotechnology.

| **SAQ 6.4** | A non-virulent retrovirus has been isolated which shows a high degree of specifity for human retina cells. This virus has been genetically manipulated and a Rb-1 gene has been inserted into its genome. Infection of cells *in vitro* with this genetically manipulated virus shows that the Rb-1 gene is expressed. Could this virus be used to prevent retinoblasloma production in children who inherit a defective Rb-1 gene? (Read our response carefully). |

| **SAQ 6.5** | Vinblastine inhibits cell division by arresting the cells at metaphase. In principle it could be used to stop cancer cells from proliferating. In practice, however, introduction of this reagent into animals also leads to the inhibition of cell division of many other types of cells. This has enormous physiological consequences. For example since the animal no longer replenishes lymphocytes and gut epithelial cells it thus is liable to infection and leakage (haemorrhaging) in the gut. Devise a strategy using a biotechnological product that might enable vinblastine to be delivered only to cancer cells. |

Summary and objectives

In this chapter we explained how malignant behaviour is a genetic property of cells. We described how the genetic differences between cancer cells and normal cells usually arise by somatic mutation, in other words, changes in genotype after the zygote stage which are not transmitted to the next generation. We also described the discovery and some properties of oncogenes and proto-oncogenes. Many mutational mechanisms such as chromosomal translocation, point mutation, deletion and insertion may account for inappropriate function of proto-oncogenes. The normal function of some of these genes is to code for proteins which regulate cell growth and differentiation. The ability to grow cells in culture has provided investigators with the opportunity to study many aspects of cancer cell biology under carefully controlled conditions. In particular, the specific contribution of a single oncogene or the effect of inactivation of a tumour suppressor gene can be readily studied *in vitro*. Although such *in vitro* assays are very useful it must be emphasised that a cautious approach should be taken, since the situation does not reflect the circumstances *in vivo* in some important respects. In the final part of the chapter, we briefly explained how the study of cancer may have an impact on biotechnology.

Now that you have completed this chapter you should be able to:

- define terms relevant to cancer: (oncogenes, transformation, tumour, neoplasm, cancer);

- describe the characteristics of the transformed phenotype (round morphology, loss of contact inhibition, proliferation in media containing low serum concentrations, anchorage-independent growth, tumorigenicity);

- describe the practical assays used to characterise the transformed phenotype; (the focus forming assay, proliferation in low serum media, growth of colonies from single cells suspended in semisolid media, subcutaneous injection into athymic mice);

- explain how tissue culture techniques have contributed to the understanding of cancer and carcinogenesis. (Isolation of oncogenes by their ability to induce the above changes in phenotype. Identification of tumour suppressor genes);

- describe the consequences of mutations in proto-oncogenes and their regulatory sequences;

- outline how biotechnology may enable the production of reagents which may be useful in cancer therapy.

Hybridomas

7.1 Introduction 134

7.2 The limitations of traditional antibody preparation 134

7.3 The basis of hybridoma technology 136

7.4 The details of hybridoma technology 138

7.5 Long term storage of hybridoma cell lines 148

7.6 Contamination 148

7.7 Hybridomas from different species 149

7.8 Human hybridomas to produce human monoclonal antibodies 149

7.9 Commercial scale production of monoclonal antibodies 149

7.10 Final comments 150

Summary and objectives 151

Hybridomas

7.1 Introduction

antibodies

antigens

Antibodies or immunoglobulins are a group of glycoproteins present in the serum and tissue fluids of all mammals. They are produced by the immune system in response to immunogenic foreign molecules called antigens and bind specifically to the antigen that induced their formation. The function of antibodies molecules is twofold, one part of the molecules bind to the antigen and the other part may bind to receptors on phagocytic cells or to the complement components which can then eliminate the antigen. Antigen-antibody reactions are highly specific. For example, antibodies produced against the measles virus will provide immunity against measles only. They will be completely ineffective against polio or mumps viruses.

use of antibodies as analytical and preparative reagents

This specificity can be exploited in the laboratory by using antibodies to detect a wide range of specific molecules. These molecules are usually, but not always, biological products such as proteins, glycoproteins, peptides, carbohydrates or nucleic acids. Antibodies may be used for both qualitative and quantitative analysis. They may also be used for topographic studies. For this, the antibodies are 'labelled' with, for example, a fluorescent dye and by measuring the presence/absence of the dye in a particular location, information concerning the distribution of antigen molecules can be obtained. Antibodies are used to label particular molecules on certain cells which can then be separated from cells not bearing these molecules (see Chapter 2). Antibodies can also be used as therapeutic agents. For example, serum containing antibodies against Hepatitis A virus may be used to treat Hepatitis A infection. Antibodies also find use in cancer therapy and forensic science. It is not surprising, therefore, that much effort is directed towards producing antibodies. This chapter is concerned with discussing the production of antibodies by *in vitro* culture.

We will begin the chapter by explaining the limitations of the traditional methods of raising antibodies in animals and the potential advantages of an *in vitro* system using hybridomas. We will then examine the details of how hybridomas may be produced and how we screen for the appropriate hybridoma. In the final part of the chapter, we will describe some aspects of quality control. The fact that we have dedicated a whole chapter to this one type of culture should emphasise to you the commercial and practical importance of hybridomas.

7.2 The limitations of traditional antibody preparation

Traditionally antibodies were produced by immunising suitable animals, for example rabbits, horses, cows, goats etc depending on how much antibody was required. The antibodies produced in this way were collected by bleeding the animal and separating the serum from the collected blood.

∏ See if you can list some disadvantages of using antibodies prepared in this way.

There are many disadvantages that you may have listed. First, the serum containing the desired antibody contains many other components which might interfere with the desired antibody-antigen reaction. Second, the animals used to produce the antibody have a finite life and each animal used will produce a different immune response to a particular antigen. Thus, a considerable effort had to be undertaken to standardise and evaluate such preparations. The most important difficulty arises, however, from the fact that sera produced from animals invariably contain many different antibodies reactive with a wide range of antigens. We call such sera, polyclonal sera as they contain many different antibodies produced by many different clones of antibody-producing cells (B cells).

polyclonal serum pool

The polyclonal serum pool of antibodies can pose problems when specific antibodies are required in high titres for particular experiments or for clinical therapy. Thus the production of homogeneous antibodies of the required specificity has been a long standing goal in immunochemical research. This goal was achieved by the development of hybridoma technology.

Homogeneous antibodies were first isolated from B cell tumours. B cells are lymphopoietic cells which differentiate in response to antigens to become plasma cells. The plasma cells secrete antibodies. Clonal populations of tumour-derived B cells can be propagated in animals as tumours or in tissue culture. All the antibodies secreted by a clone of B cells are identical, so these tumour cells provide a good source of homogeneous antibody. Unfortunately it is not easy to produce B cell tumours which provide antibodies of the desired specificity. Plasma cells not derived from tumours cannot be cultured *in vitro*, so they cannot be used as sources of homogeneous antibody.

monoclonal antibodies

In 1975 Kohler and Milstein developed a technique by which clonal populations of cells secreting antibodies of defined specificity could be fused with myeloma (B cell tumour) cells to produce immortalised cell lines which continued to produce homogeneous antibody. These hybrid cell or hybridomas can be propagated *in vitro* and can be cloned to produce antibodies against specific antigens. The antibodies produced by hybridoma clones are called monoclonal antibodies.

∏ Suggest some advantages and disadvantages associated with producing monoclonal antibodies instead of polyclonal serum antibodies.

The main advantages of producing monoclonal antibodies using immortalised hybridoma cell lines are:

* we are able to produce pure preparations of antibodies with known specificities (ie against defined target molecules);

* we can produce standardised antibody preparations whose properties are the same (or very similar) from batch to batch. Remember that by using animals, production is discontinuous and on the death of the animal, the antibody preparation made via replacement animals are often quite different because of the genetic and physiological variations found amongst animal populations. In principle, by using immortalised hybridoma cell lines we can produce unlimited amounts of particular antibodies;

* we are able to use relatively impure antigens to raise specific antibodies. This is, as we shall see, because we select particular clones of antibody-producing cells during the production of hybridoma cell lines. Thus, although a complex mixture of antigens may be used initially, the ultimate product is a pure antibody preparation.

There are, however, some disadvantages of producing monoclonal antibodies.

antigenic shift
(immunological
drift)

Monoclonal antibodies are very time consuming and labour intensive to produce. They require considerable technical expertise to produce and screen and are not always the best choice for many immunochemical reactions. There are some situations where a high degree of specificity compromises sensitivity in a reaction. For example, many viruses alter their surface antigens continuously in order to escape the host's immune response. This phenomenon is called antigenic shift or immunological drift. The host produces a new antibody response to each variant as it arises so the serum of the host will contain antibodies to a number of different variants of the same virus. Such a polyclonal serum is very much more useful for detecting the presence of that virus in other patients during a virus epidemic than a monoclonal antibody which can only detect one variant which may or may not be present in other patients.

Having established a case for producing monoclonal antibodies, let us now turn our attention to how to achieve their production.

SAQ 7.1

Hepatitis B virus (HBV) is known to produce a number of surface antigens. These are designated as HBs antigens. Analysis has shown that the most common subtypes of HBs antigens are adr, ayw and adw. (Note that a, d, r, w, y are epitopes which react with specific antibodies).

Which of the following approaches is most likely to result in a satisfactory antibody-based procedure for detecting Hepatitis B infections in a population? (Give reasons).

1)   Antiserum raised in goats against intact viruses.

2)   Monoclonal antibodies produced by hybridoma technology against epitope w.

3)   Monoclonal antibodies produced by hybridoma technology against epitope a.

7.3 The basis of hybridoma technology

Hybridomas are cells produced by the fusion of two types of somatic cells leading to the production of cells capable of producing antibodies. Kohler and Milstein showed that antibody-secreting cell lines could be propagated in tissue culture by fusing short lived antibody-producing B cells with tumour cells capable of continuous cultivation *in vitro* (that is, immortal cell lines). Thus the whole basis of hybridoma technology is to combine the growth characteristics of tumour cells with the antibody-producing capabilities of particular B cells.

There have been few changes to the strategy and procedures developed by Kohler and Milstein and hybridomas are still routinely produced by fusing B cells with myeloma cells. You should note that much of the early studies were done using cells derived from mice. In many instances these are still the cells that are employed although there is a progressive development of hybridoma technology using cells from other species. Irrespective of species, the general strategy that is employed to produce monoclonal antibodies is similar. Here we will predominantly focus on monoclonal antibody production using cells derived from mice.

$\prod$ See if you can identify the three major problems that need to be solved in order to produce a desired hybridoma cell line.

The three major problems that have to be solved in producing hybridomas are:

- the best fusion partners have to be found;

- ideal conditions for the fusion have to be defined;

- an appropriate screening system has to be developed in order to identify the desired hybridoma cell lines.

We will deal with the latter two items in section 7.4.

7.3.1 Choosing the fusion partners

MOPC

Myelomas can be induced in some strains of mice by injecting mineral oil into the peritoneum of the animal. In 1972 Polter isolated some myelomas from the BALB/C strain and this strain is still the most commonly used in hybridoma production. Tumour cells derived by this method are known as MOPC (mineral oil plasmacytomas). Myelomas have all the cellular machinery needed for producing immunoglobulins and many secrete more than one type of antibody. Myelomas that are to be used for fusion need to be selected so that they do not, by themselves, produce functional antibodies.

splenocytes are used

The antibody secreting cells are isolated from immunised animals, usually mice or rats. These cells must carry the immunoglobulin genes which specify the desired antibody. The antibody secreting cells used are usually splenocytes (ie antibody secreting spleen cells).

Hybridomas can be made by fusing cells from different species, but the efficiency of fusion is very much lower than when cells from the same species are fused. Cells from all strains of mice can be used to fuse with BALB/C myelomas, but usually BALB/C immunised splenocytes are used so that the hybridomas can be grown as tumours in this strain without raising an immune response.

SAQ 7.2

Why are cells from spleens used to produce hybridomas rather than cells from other organs? Suggest other organs that might be used.

7.3.2 Steps in producing hybridomas

Four basic steps are involved in producing hybridomas and monoclonal antibodies:

- immunising mice;

- fusing cells;

- (developing the) screening procedures;

- producing the hybridomas.

7.4 The details of hybridoma technology

7.4.1 Immunising mice

∏ Why are animals immunised before collecting their spleens for the preparation of B-cells?

Your ability to answer this question will depend upon your previous experience of immunology. When an antigen is introduced into an animal, it stimulates the B cells, which produce antibodies which react with the antigen, to grow and divide. This process is quite complex and results in the clonal expansion of certain B cells (note that details of this process are given in the BIOTOL text 'Cellular Interactions and Immunobiology). Thus by immunising an animal, we increase the proportion of B-cells capable of producing the desired antibody (ies). This cellular response is reflected in the level (titre) of these antibodies in the blood of the immunised animal (Figure 7.1). This antibody response matures as the animal is repeatedly exposed to the same antigen (Ag). The degree of response (in terms of B cell numbers and antibody production) is dependent upon the type of antigen, dose size, number of exposures and the animal's tolerance to the antigen. Since the physical nature of the antigen is particularly important, we will briefly examine this aspect.

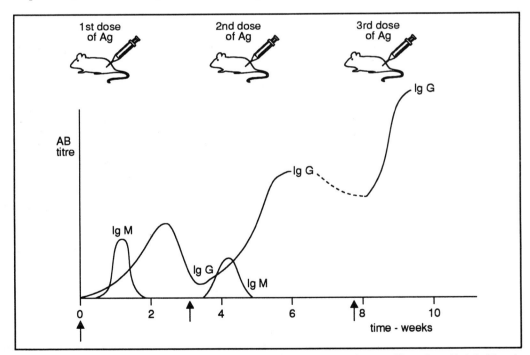

Figure 7.1 Boosting the antibody response by multiple immunisations with a specific antigen (Ag). Ig M and Ig G are two different classes of antibodies. Arrows indicate inoculation time.

Types of antigen

Before we examine the various types of antigens we need to distinguish some terms.

antigen
immunogen

An antigen is a compound which will react with an antibody. An immunogen is a compound which will stimulate the production of antibody(ies). Most antigens are also immunogens and often within the literature the two terms are used interchangeably.

hapten

A hapten is a compound which by itself will not stimulate antibody production, but in the presence of a carrier will become immunogenic. The antibodies produced by such a stimulus will react with the hapten.

epitope

An epitope is the part of an antigen which reacts with a particular antibody. Thus large complex antigens may contain many different epitopes each of which will react with different antibodies.

SAQ 7.3

Dinitrophenol has the following structure:

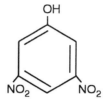

When we combine this compound with a carrier protein (for example bovine serum albumin) and inject it into a mouse, the mouse produces antibodies, one type of which react with dinitrophenol and several other types of which react with bovine serum albumin.

If, however, dinitrophenol itself is injected into a mouse, no antibodies are produced. On the other hand injection of bovine serum albumin into mice leads to the production of several different antibodies with quite different specificities.

1) What term best describes dinitrophenol? (Choice: hapten, immunogen).

2) What term best describes bovine serum albumin. (Choice: immunogen, hapten, antigen).

3) Can dinitrophenol be described as bearing a single epitope or multiple epitopes?

4) Does bovine serum albumin carry a single epitope or multiple epitopes?

---

soluble proteins

Soluble proteins produce good responses with doses as low as 1mg/injection (that is they are highly immunogenic). Usually 10-50 mg/injection dose of soluble protein are used for immunising mice. Adjuvants or carrier proteins such as haemocyanin can be used to enhance the immune response.

particulate
proteins

Particulate proteins in suspension also make good immunogens because they are readily phagocytosed by antigen-presenting cells and this is necessary to stimulate B cell proliferation.

synthetic peptides
Synthetic peptides are usually administered with carrier proteins such as bovine serum albumin (BSA) or keyhole limpet haemocyanin (KLH) (ie as peptide: carrier complexes). Good responses to these peptides are found and the monoclonal antibodies produced with synthetic peptide: carrier complexes are very highly specific for small molecules or particular epitopes.

cells as immunigens
Live, attenuated or killed cells can also be used as immunogens but the antibody produced is usually of low affinity. Bacteria and yeasts can be used but in this case great care should especially be taken not to spread infection into the animal house.

nucleic acid
Nucleic acids are weak antigens and have to be administered with carrier proteins.

carbohydrates
Simple carbohydrates are poor immunogens, but complex carbohydrates and polysaccharides can be used when coupled with carrier proteins.

Route of inoculation

The easiest way to inject a mouse is intraperitoneally or intramuscularly.

Intravenous (i/v), subcutaneous (s/c) or intradermal (i/d) routes can also be used, but the responses are produced more slowly. Intraperitoneal injection is particularly useful for administrating particulate protein immunogens.

Usually female mice are used because they are easier to handle than males. More than one animal are often immunised because no two animals produce the same response to the same antigen.

Test bleeds

A single immunisation event will not produce the desired type or level of antibody response. Fusion must not be started until serum from the test bleeds show antibodies of the titre and specificity required. Hyperimmunisation, (ie multiple injections) with the same antigen over a period of several weeks, will produce antibodies of higher affinity for the antigen.

Tail bleeds are collected periodically after immunisation and checked for the presence and titre of the desired antibody. As the immune response matures increasing levels of specific antibody will be found in the test sera. Care must be taken to ensure that high levels of other antibodies are not also present - these are usually the result of contaminants in the injected preparation or infectious agents in the mouse's environment. If the mouse shows signs of illness it must be isolated and not used for hybridoma production until it has recovered.

7.4.2 Developing the screening procedures

setting up screening procedures before cell fusion
Because most hybridomas grow at approximately the same rate, the tissue culture supernatants containing antibodies all become ready to screen at once. The screening process is very labour intensive, so carefully chosen assays are essential and must be set up before the hybridomas are made. There is no time to carry out any fine tuning of screening assays once the hybridomas have started growing. Sera from the test bleeds may be used to set up and validate the screening tests.

About one week after the fusion, colonies of hybrid cells are ready to be screened. Samples of the tissue culture medium are removed from wells containing growing hybridomas (we will describe these in more detail later). Culture fluids, (CF), are screened for the presence of antibody. Successful fusions normally yield between 500-1000 hybridoma colonies, sometimes more. All of these colonies need to be screened over a two to six day period, so the ideal screening test must allow cultures to be screened in batches of 50 or so at a time and be able to identify specific antibody producing cells in less than 48 hours.

Three types of screening methods are routinely used for checking hybridomas:

  a) antibody capture assays.

  b) antigen capture assays.

  c) functional assays.

We will outline these assays here but, if you wish to follow this up in more detail we would recommend the BIOTOL text 'Technological Application of Immunochemicals'.

a) Antibody capture assay

These are probably the most convenient assays for hybridoma screening. Antigen is bound to a solid substratum such as the base of a 96 well microtitre plate or allowed to coat the surface of plastic beads. The culture fluid to be tested is exposed to the antigen and allowed to bind for anything from 1 hour to overnight. Any antibody in the culture fluid which binds to the antigen remains attached to the solid matrix as antigen-antibody complexes, while unbound antibody can be washed off. A second antibody which reacts with the antigen-antibody complex is then added. This second antibody is conjugated with either an enzyme which produces a colour change in the presence of an appropriate substrate or with a fluorochrome. This reaction is allowed to proceed for 20-60 minutes and then any second antibody that is unbound is washed off. The presence of the first antibody (ie from the culture fluid) is detected by adding substrate or by examining the plate with a UV reader. We have represented the process in Figure 7.2a.

This is a quick and easy assay, but is dependent on the availability of pure antigen for coating the matrix. Whole cells which carry the specific antigen on their surface may also be used to bind antibody in the same way.

b) Antigen capture assay

This procedure can either be operated in solution or by allowing the antibody to coat a microtitre well or beads. We will describe the second version of this type of procedure.

The culture fluid from the hybridoma culture is incubated in a microtitre well where antibodies are allowed to attach to the well walls. The culture fluid is then removed and the wells are washed. A solution or suspension of the antigen is then introduced into the well. After incubation, the solution is removed and either the amount of antigen remaining in the well or the amount left in the solution is measured. In this way culture fluids containing the appropriate antibody can be detected (see Figure 7.2b).

This method is quite fast, but is only really useful for detecting very high affinity antibodies.

Since it is difficult to ensure that all of the antibody binds to the solid matrix, it is often preferable to use antibody and antigen in solution. In this case, however, we have to have a simple and sensitive technique for distinguishing between bound and unbound antigen. If the antigen is very small compared to the antibody, it is usually quite easy to separate bound and unbound antigen by gel filtration or by adsorption onto activated charcoal.

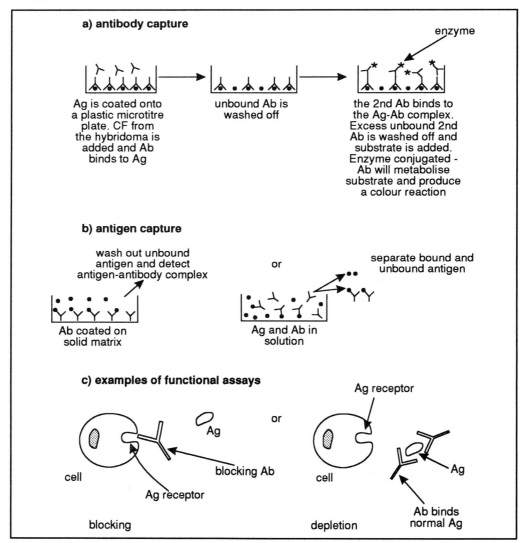

Figure 7.2 Antibody screening assays. a) antibody capture, b) antigen capture and c) functional assays. Ag = antigen; Ab = antibody; CF = culture fluid. See text for details.

Let us assume that you wish to raise a monoclonal antibody against oestrogen. Oestrogen by itself will not stimulate antibody production but if attached to bovine serum albumin (BSA) it will (in other words, oestrogen is a hapten). A mouse has been immunised with oestrogen-BSA conjugate and has been shown to produce antibodies which react with oestrogen. B cells from the spleen of this mouse have been collected and fused with myeloma cells. The fused products have been inoculated at low density into small aliquots of culture medium and incubated. You are now attempting to identify which aliquots of culture medium contain hybridoma cells which will produce anti-oestrogen antibodies.

You have available radioactively labelled (^{3}H) oestrogen. Suggest a possible screening procedure.

c) Functional assays

As the name suggests, the antibodies in the culture fluid are assessed by their effect on a function or activity of cells. Usually this involves blocking a reaction by making an essential binding site unavailable or by depleting an essential component of a reaction mix by precipitation of an antigen-antibody complex.

For example, in the cases shown in Figure 7.2c, the antibody either blocks the binding of the antigen by the cell or removes the antigen. In either case, the cell will not respond to the same extent as it would to the antigen in the absence of the antibody. This is the basis for detecting the antibody.

7.4.3 Making hybridomas

Once a good immune response has been obtained in the animal and the screening assays are set up and ready to use, the B cells and the myeloma cell can be fused. Antibody secreting cells are extracted by dissecting out the spleen of the immunised animals and cutting up the tissue to form a single cell suspension. The myeloma cells should be extracted from animals injected with mineral oil several days before the fusion experiment and the cells should be maintained in tissue culture or, alternatively, the myeloma cells could be thawed from liquid nitrogen stocks 2-3 days prior to use. It is important to check their viability and ensure that they are free of microbial contaminants. The myeloma cells are sub-cultured one day before the fusion and fed with medium containing 10% serum. The cell density should be adjusted to give a concentration of 5×10^5 cells ml^{-1}. Immediately before fusion the cells are diluted with an equal volume of medium supplemented with 20% foetal calf serum and a double strength OPI (oxaloacetate, pyruvate and insulin) solution. The OPI solution provides essential nutrients and hormones for growing cells at low densities.

preparation of cells for fusion

use of polyethylene glycol (PEG)

The most commonly used agent for fusing myeloma and antibody secreting cells is polyethylene glycol (PEG). PEG is a polymer of ethylene glycol and comes in different molecular weight sizes from PEG 1000 to PEG 6000. For most hybridoma fusions, PEG 1500 is used and it must be of high quality. PEG works by fusing the plasma membranes of adjacent myeloma and/or antibody-secreting cells to form a single cell with two or more nuclei (called a heterokaryon). The heterokaryon retains these nuclei until the nuclear membranes dissolve prior to mitosis. During mitosis and through further rounds of division the individual chromosomes segregate into the daughter cells. The abnormal number of chromosomes in the fused cell means that segregation does not always result in identical sets of chromosomes being delivered to each of the daughter cells. Some chromosomes may be lost. If one of the chromosomes carrying a functional, rearranged immunoglobulin heavy or light chain gene is lost, antibody production will stop in that daughter cell. If these daughter cells are viable and divide to produce a hybridoma together with other daughter cells which can secrete antibody the amount of antibody produced will be decreased.

∏ How would you select out the hybridoma cells which are secreting the antibody you require?

You can compare your ideas with this procedure described in the next section.

7.4.4 Elimination of unfused cells by drug selection

Only about 1% of the starting cells actually fuse and only about one fusion in 10^5 forms a viable hybridoma even in the most efficient fusions. So you can see that this would leave a lot of unfused cells still in the culture. The plasma cells from the immunised animals do not grow in tissue culture and will die away fairly rapidly. Myeloma cells, however, are very well adapted to *in vitro* culture and will continue to proliferate and overgrow the hybridoma cells unless they are removed.

salvage
pathway of
nucleotide
synthesis

Usually the myeloma cells used have a mutation of one of the enzymes of the salvage pathway of purine nucleotide synthesis. The mutant gene(s) is carried on the same chromosome as the immunoglobulin genes. If a drug is incorporated into the culture medium which blocks the normal purine synthesis pathway in these cells (for example, 8-azaguanine or aminopterin) the myeloma cells which cannot use the salvage pathway are killed because they cannot synthesise purine. The gene that myeloma cells lack codes for the hypoxanthine guanine phosphoribosyl transferase (HRPT) enzyme which catalyses the salvage pathway for purine nucleotide synthesis. Unfused cells and fused cells without the required chromosome will be HRPT⁻ and have only the *de novo* synthesis pathway (which is blocked by the drug) by which to synthesise purines. This type of selection is known as HRPT or HAT selection.

HAT selection

HAT stands for Hypoxanthine Aminopterin and Thymidine. The basis of HAT selection is, therefore, that it only allows the growth of cells which have a functional salvage pathway for purine biosynthesis. We have represented this diagrammatically in Figure 7.3.

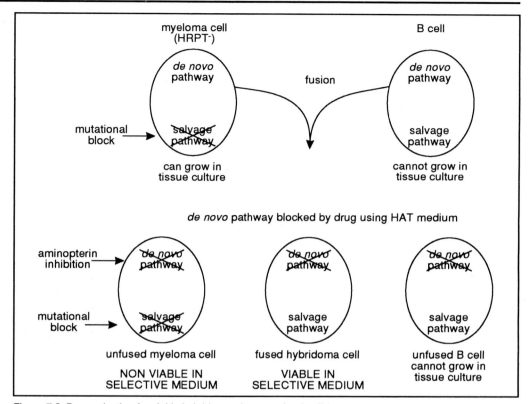

Figure 7.3 Drug selection for viable hybridomas (see text for details).

The drugs methotrexate, azaserine or 8-azaguanine can be used in place of aminopterin.

Now that we have established the basic steps in producing hybridomas, let us return to the key step of cell fusion and examine the details of this step. Several methods can be used but here we will confine ourselves to the two principle methods; the so called stirring and the spinning methods.

7.4.5 Two methods of fusing cells

Stirring method for fusing cells

In the stirring method, splenocytes and myeloma cells are washed separately by centrifuging at 400x g in serum-free medium and resuspending in 1-5 ml of serum-free medium. The cells are mixed and centrifuged to get the cells into close proximity to each other. All of the medium is then carefully pipetted off and 1 ml of a 50% solution of PEG 1500, which has been prewarmed to 37°C, is added very slowly to the cell pellet while gently stirring to resuspend the cells. Then 1 ml of serum-free medium is added dropwise over a period of about one minute, while constantly stirring the cell suspension. A further 9 ml of serum free medium are then added over a period of two minutes, still stirring continuously.

The cells are centrifuged again at 400x g for 5 minutes, the supernatant is removed and the pellet resuspended in 10 ml of prewarmed medium with 20% foetal calf serum, OPI solution and selection medium containing aminopterin and hypoxanthine and thymidine (HAT) or 8-azaguanine. The suspension is made up to 200 ml with medium aliquoted into each well of 20 x 96 well plates and incubated at $37^{\circ}C$ in a CO_2 incubator. The cells are examined using a microscope after day 2 and are visible to the naked eye by days 4-6. (You might like to draw yourself a flow diagram of these steps).

Spinning method for fusing cells

Cells can also be fused using the spinning method, which requires a lower concentration of PEG. The cells are centrifuged, washed and mixed, then recentrifuged as for the stirring method above, and then the pellet resuspended in 0.2 ml of a 30% PEG solution at $37^{\circ}C$. The cells are resuspended by gently tapping the base of the centrifuge tube. They should not be pipetted. After centrifuging again at 400x g for 5 minutes, the cells are resuspended in 5 ml of serum-free medium. Again, the tubes are tapped or flicked to resuspend the cells. A further 5 ml of medium supplemented with 20% foetal calf serum, OPI and HAT as above are added. The cells are mixed gently and made up to a volume of 200 ml and aliquoted into 96 well plates as before.

There appear to be no particular advantages or disadvantages with either method, and the choice of method depends entirely on convenience.

7.4.6 Screening hybridomas

Wells containing growing hybridomas are ready to screen from 7-14 days after fusion, or when the colonies become just visible to the naked eye. 50 µl of the culture supernatant is removed aseptically from the top of the well, taking care not to disturb the cells at the bottom. 50 µl of fresh medium is added to the wells and the hybridomas are re-incubated. The supernatant can be screened using any of the screening procedures described earlier.

If no samples are found to be positive for the antibodies required even though a strong immune response was produced in the animal, then the fusion is repeated and a different screening method is employed. If a weak positive result is produced and the immunisation response was also weak, the immunisation procedure should be re-evaluated.

The wells that do score as antibody-positive should be fully tested as soon as possible with the assay system in which they are to be used. If a range of monoclonal antibodies is required further fusions may need to be set up and screened.

7.4.7 Cloning hybridomas

∏ Assume that you have identified a well which contains hybridoma cells which are producing the desired antibody. Is such a culture a pure culture (that is containing only one cell type)?

The answer is probably no. The well from which the supernatant was taken and tested will contain more than one hybridoma clone, together with several cells with an undesirable assortment of chromosomes. Single cell cloning is essential to produce a pure culture but it is a difficult process requiring a great deal of time.

single cell picking

The initial cloning is done using the limiting dilution method described in Chapter 2. Once these clones are established they are screened for antibody production and the positive clones grown up. A second cloning step, known as single cell picking is used to pick off individual cells using a capillary tube and a microscope. Hybridoma cells have very low plating efficiencies, so single cell cloning is usually done in the presence of feeder cells or using conditional medium. You may like to refresh your memory of these procedures by referring to the relevant section of Chapter 2.

Cloning on feeder cells

Feeder cells provide the growth conditions that cells need if they are to be grown at very low densities. Feeder cultures may be prepared from macrophages, spleen cells, thymus cells or fibroblasts. If the feeder cells are pretreated with mitomycin C or lethally irradiated once they are confluent, they provide ideal growth conditions for cloning without contaminating the hybridomas. Fibroblast feeder cultures (the MRC5 cell lines from the American Type Culture Collection are the best) can also be used.

Cloning in soft agar

This is the most popular method for single cell cloning of hybridomas. It does have the disadvantages of being slow to perform and being very selective for the type of cell that will grow in soft agar. Most hybridomas do, however, grow well in agar. Cells are diluted as for the limited dilution cloning method giving between 10^5 and 10^2 cells per ml. These are then dispensed in 100-200 ml aliquots of double strength medium, into the wells of 24 or 6 well tissue culture plates. The cultures are overlaid with a 1.5% solution of warmed agar which is allowed to set in the refrigerator before incubating the plates of $37^{\circ}C$ in a CO_2 incubator. Colonies visible to the naked eye will appear at around day 10 and can be picked off within a plug of agar. The cells are dispersed by gentle pipetting and warming in medium. After washing, the cells are resuspended in medium supplemented with 10% foetal calf serum and selection medium before incubating at $37^{\circ}C$ in CO_2 for 48-72 hours. Supernatants are screened for antibody as before and any clones which are producing good levels (titres) of specific antibodies can be expanded by sub-culturing. If the cloned hybridomas fail to produce antibody or if the antibody is of low titre further single cell clones may need to be selected in order to find a stable clone that produces a high titre of specific antibody. It is important to re-test the specificity of the monoclonal antibody at this stage.

SAQ 7.5	Assume that a mouse produces 10^5 different types of B cells. (Note that each type of B-cell produces its own specific antibody). If such a mouse is immunised with antigen A, the B cells which produce antibodies which react with antigen A proliferate such that they make up about 0.1% of the B-cell population of the spleen. About 10^{10} splenic B cells have been isolated and subjected to fusion with myeloma cells. About 1% of these B-cells successfully fuse with myeloma cells but only 0.01% of the fused cells remain viable. What is the probable number of these viable hybridomas that will produce antibodies which will react with antigen A?

7.4.8 Sub-culturing hybridomas

Hybridomas, like myelomas are easy to culture *in vitro*. The cells grow in suspension and do not readily adhere to plastic or glass. Hybridoma clones are usually sub-cultured when they have grown to a density of approximately 10^6 cells ml^{-1} provided the cells appear healthy under the microscope. Healthy cells are refractile (glassy) and have smooth membranes, whereas cells which are distressed and dying are granular and dark, with rough membranes. It is not necessary to count the cells before sub-culturing if the cells are healthy and growing well. Hybridoma cultures may be split or divided 1 to 10 or even 1 to 20 depending on their vigour. Cells are allowed to settle and most of the culture fluid is drawn off. The cells are suspended in a 10-20 ml of fresh supplemented medium and 1 ml aliquots of the suspension are used to seed 10-20 new flasks or tissue culture plates. A further 5-10 mls of fresh medium are added to each flask or plate.

Antibody production should be checked at this stage and the cultures expanded by sub-culturing further to give stocks for storage. A lot of monoclonal antibody production work is done commercially in large stirred reactors which work in 1000 litre volumes of hybridoma cultures.

7.5 Long term storage of hybridoma cell lines

Π Without looking back to earlier chapters, describe how you would preserve your hybridoma clones once they have been established and screened. Pay particular attention to preserving viability and avoiding contamination. How would you mark your cultures for identification?

Large stocks of hybridomas should be stored in liquid nitrogen, usually in batches of 20-50 vials at a time, as an insurance in case working stocks are, or become, contaminated. At this stage you may also need to consider patenting your clone or the monoclonal antibody it secretes. Once a patent has been applied for, the clones may be banked with a cell bank or B type culture collection. Hybridoma cells are frozen in the same way as other cells, using a cryoprotectant such as DMSO and cooling at 1°C per minute down to -70°C before transferring to liquid nitrogen.

7.6 Contamination

Making and screening hybridomas is very hard work and requires a considerable investment of time. It is rare that hybridomas secreting high affinity, specific monoclonal antibodies, can be produced in less than 6 months, a year or two is more usual. It is vital, then, to ensure that contamination by bacteria, fungi or mycoplasmas is carefully avoided, particularly in the early days after fusion when valuable clones can be lost. Aseptic technique must be vigorously applied. Expanded clones for storage must be checked for contamination before stocks are frozen down. If contamination occurs later on in the hybridoma cultures, the clone can be discarded and a frozen vial thawed to replace it.

We have discussed these issues elsewhere but it is so important that we need to re-emphasise them here.

mycoplasma detected by using Hoechst Dye, DNA hydrisation or by monoclonal antibodies

Bacterial and fungal contaminations are relatively easy to detect and eradicate. Mycoplasmas are much more difficult to control and require continuous vigilance to protect the cultures from infection. Mycoplasmas do not affect the turbidity of the culture medium and are difficult to detect by microscopy because they do not grow in large numbers. Like all cell lines, hybridomas should be routinely screened for mycoplasma infection. DNA hybridisation and antimycoplasma monoclonal antibody-based kits are available commercially. The most commonly used method in routine cell culture laboratories is the Hoechst Dye (33258) staining assay. A sample of cells is fixed with methanol and stained with Hoechst Dye. Infected cells are identified by the presence of the DNA stain in the cytoplasm of the cells (cellular DNA remains in the nucleus except during mitosis). Antibiotics such as lincomycin and tylosin do have some activity against mycoplasmas, but the organisms are so prevalent that antibiotic resistance soon becomes a problem if these agents are used regularly. As far as contamination with any organisms (especially mycoplasma) is concerned, prevention is always better than cure. Ensuring that all media and media supplements are pre-screened for mycoplasmas goes a long way towards protecting valuable cultures from mycoplasma infection.

7.7 Hybridomas from different species

interspecies hybridomas

Antibody secreting cells from one species can be fused with myeloma cells from another species to produce interspecies hybridomas. Rat B cells may be fused with mouse myeloma cells to yield hybridomas producing rat antibodies. Interspecies hybridomas are very difficult to produce and are of limited use in the production of good monoclonal antibodies. They are rarely used nowadays.

7.8 Human hybridomas to produce human monoclonal antibodies

The production of human monoclonal antibodies is a very exciting new application of hybridoma technology, particularly because of the potential value of monoclonal antibodies in clinical therapy. For instance, if highly specific monoclonal antibodies can be made against tumours these would reduce the need for whole body radiation or toxic chemotherapy in the treatment of cancer.

∏   Suggest why it is not possible to produce human monoclonal antibodies as easily as it is to produce mouse or rat monoclonal antibodies.

use of EBV to transform human B cells

Naturally, it is not ethical to induce myelomas in human, nor is it possible to immunise humans in the same way as animals. Human subjects who have already encountered certain antigens may be used as sources of antibody secreting cells but it is difficult to find suitable myeloma fusion partners. However, B cells from humans can be separated quite easily from small blood samples. These cells may be transformed *in vitro* using the Epstein Barr virus (EBV) to produce immortalised B cells which can replicate and grow in tissue culture. Unfortunately EBV transformed clones by themselves do not produce very large amounts of immunoglobulins, but recent research on the fusion of EBV transformed B cells with mouse myeloma cells has produced promising results.

7.9 Commercial scale production of monoclonal antibodies

use of bioreactors

Once suitable hybridomas have been produced, it is essential to scale up the production of antibodies to achieve an appropriate yield. Two basic and quite different strategies can be used. One is to use large scale vessels (bioreactors) in a manner analogous to the procedures used in fermentation technology to produce such desirable compounds as antibiotics from microbial cell culture. We have described such vessels elsewhere in this text (Chapter 8) so we will not elaborate further. Yields are generally fairly low (in terms of $\mu g \ ml^{-1}$) but, of course, the culture conditions are easily controlled and maintained, yields are fairly predictable, downstream processing fairly straightforward and a standardised product is made. The process does, however, require expensive bioreactors and is, therefore, capital intensive.

An alternative to *in vitro* cultivation is to introduce the hybridoma cells into the peritoneal cavities of suitable animals. Here the cells grow and produce ascites tumours. The serum from such animals contains high levels (mg ml^{-1}) of the antibody produced by the hybridoma cells.

∏ It might seem strange to go to all the trouble of producing cloned hybridoma cells *in vitro* to avoid the problems of producing polyclonal antibodies using whole animals only to re-introduce these cells into an animal. However, it is quite logical. See if you can identify the rationale of this strategy.

the ratio of the concentrations ascitis-derived antibody and other antibodies is important

The first point to make is that raising antibodies using the ascites route is relatively cheap. Because the yield of antibodies is very high (mg ml^{-1} range) using ascites compared to the μg ml^{-1} range produced in bioreactors and no expensive capital equipment is required, the ascites route may be particularly attractive. Of course, the antiserum containing the monoclonal antibody produced via the ascites tumour route will be contaminated by antibodies produced by the host animal's normal B cells. However, these antibodies will invariably be at much lower concentration than the hybridoma-derived antibodies. Dilution of sera derived from such animals to give the appropriate working concentrations of the desired antibodies means that the concentrations of other antibodies and contaminating serum constituents are negligible. The use (or as some would claim, abuse) of animals simply as protein factories to generate particular proteins is, however, argued by many as unethical. With improvements in animal cell technology the general move is towards producing monoclonal antibodies in bioreactors rather than from ascites tumours. Especially important has been the development of *in vitro* dialysis (membrane) reactors which enable the continuous separation of cells and their excreted protein products. Although these still present a number of problems (for example clogging of the membranes), yields can be improved.

7.10 Final comments

We began this chapter by indicating that the practical use of antibodies is very extensive. It pervades many activities in a wide variety of sectors. The advantages of monoclonal antibodies implies that these are being widely adopted as therapeutic and diagnostic agents in medicine, as quality control agents to detect microbiological and chemical contamination in a wide range of medical and food products, as analytic reagents in the measurement of numerous compounds, as reagents to help cell identification and sorting and so on. The production of monoclonal antibodies has become very big business indeed. All the indications are that the business activity will continue to expand and we can anticipate further extension of the already long list of products of immunochemical suppliers.

electrofusion may be an alternative to PEG treatment

We might, however, also anticipate that some changes may occur to improve and extend current practices. For example the use of high voltage electric pulses (electrofusion) instead of PEG treatment to fuse cells seems to offer some promise in improving fusion efficiency. Perhaps, however, the most important advances which may be made will involve the use of recombinant retroviruses as genetic vectors. These may be used to deliver oncogenes (cancer-inducing genes) into antibody-secreting cells, thereby immortalising them without the need for fusion with myeloma cells. Antibody-secreting cells would still need to be screened and identified. We will discuss the use of retroviruses as genetic vectors for use with animal cells in a later chapter.

Summary and objectives

In this chapter we have discussed the potential advantages of producing monoclonal rather than polyclonal antibodies. We also explained how the production of monoclonal antibodies depends upon our ability to fuse B cells with myeloma cells and to screen and identify suitable hybridoma products. We briefly described how such cell lines may be used for large scale *in vitro* cultivation techniques or to induce ascites tumours in appropriate host animals.

Now that you have completed this chapter you should be able to:

- compare the advantages and disadvantages of using monoclonal and polyclonal antibodies;

- discuss the need for monoclonal antibodies as research tools and their potential use in clinical therapy, diagnosis and analysis;

- explain how to immunise animals and how to produce myeloma cells for fusion to form hybridomas;

- describe how to select fused, antibody producing cells and how to screen hybridomas that are likely to yield useful antibody;

- describe cloning procedures and the steps involved in the production of monoclonal antibodies;

- explain why the production of monoclonal antibodies represents such a large part of the commercial application of *in vitro* animal cell culture.

Large scale animal cell culture

8.1 Introduction	154
8.2 Culture parameters	154
8.3 Scale up of anchorage-dependent cells	157
8.4 Culture vessels	158
8.5 Suspension culture	167
8.6 Increasing cell densities and continuous flow cultivation	173
8.7 Selecting the appropriate bioreactor system	178
8.8 Applications of large-scale animal cell cultures	180
8.9 Regulatory issues	182
Summary and objectives	184

Large scale animal cell culture

8.1 Introduction

Routine cell culture for the purposes of studying cell morphology and function or toxicity and drug efficacy testing, can be performed on a fairly small scale in most laboratories. Culture flasks of up to 175 cm^2 surface area are ideal for establishing new cell lines and provide up to 1×10^7 cells per flask. However, in order to provide sufficient cell numbers for the production of vaccines, antibodies, growth factors or hormones it is not feasible simply to multiply the number of flasks by a hundred or thousand fold. For these purposes, it is necessary to scale-up the whole culture from a 1 litre system to a 100 litre or 1000 litre system or greater. Increasing the size of the culture vessel alone is not sufficient; a number of parameters need to be adjusted to provide an adequate growth environment for large scale culture. Financial investment can be substantial: whereas with small scale culture the loss of a few flasks of cells through contamination or nutrient deficiency is inconvenient, with cultures of 2-3000 litres a single contamination event can mean the loss of a great deal of time and money.

In this chapter, we will look at some of the adjustments that need to be made to scale up both suspension and monolayer cultures and discuss some uses for large scale production of animal cell lines. (Note that several BIOTOL text are devoted to the physical processes, design and operation of large scale bioreactors. If you wish to follow up these aspects we recommend 'Operational Modes of Bioreactors' and 'Bioreactor Design and Product Yield').

∏ From your knowledge of cell cultures, make a list of (micro-) environmental factors that need to be taken into consideration when scaling up of animal cell lines. (You will be able to check your list with the topics covered in the following sections).

8.2 Culture parameters

Here we will divide our discussion into:

- medium;

- non-nutritional medium supplements;

- pH;

- oxygen;

- redox potential.

8.2.1 Medium

replenishment
of glutamine

High cell densities and long term culture can result in rapid depletion of nutrients from the culture medium. Minimum Essential Medium (MEM) or Eagle's Basal Medium supplemented with serum may be adequate for small scale culture, but for scale up it is necessary to use an appropriate complete medium with serum or serum substitute, plus either peptone or bovine serum albumin. Depletion of glutamine can be growth limiting quite early on in the culture. Glutamine is converted to glutamic acid, leucine and isoleucine by cellular and serum enzymes and, therefore, needs to be replenished frequently. Human diploid cells have a high requirement for cystine. Depletion of nutrients can be growth limiting long before the nutrient actually runs out. Cells tend to pack tightly together as they begin to 'starve', so the surface area available for nutrient uptake becomes smaller, resulting in the slowing down of growth and eventually in cell death.

batch feeding
of glucose

Glucose metabolism and oxidative phosphorylation are both involved in the production of ATP. By-products of glucose metabolism, in the presence of excess glucose, are lactic and pyruvic acid which lower the pH of the medium, so it is important to regularly supplement the medium with glucose in small amounts (ie batch feed) rather than to add all of the glucose at the beginning of the culture.

perfusion
cultures

Essential amino and fatty acids, hormones, vitamins and growth factors also need to be replenished regularly during the life of the culture. In static or 'batch' cultures where there is a set volume of medium in a closed system, it is necessary to decant off some or all of the 'spent' medium and replace it with 'fresh' medium at regular intervals. The operation is made simpler with cultures that are maintained by perfusing medium through the culture so that medium enters at a set rate and is removed at the same rate. The advantages and disadvantages of both systems are discussed later (section 8.5).

8.2.2 Non-nutritional medium supplements

Large scale cultures which are grown in deep tanks need to be stirred and aerated continuously in order to maintain a homogenous environment. Mechanical stirrers can set up shear forces which damage cells and also foaming, caused by aeration, damages cells as a consequence of the interaction of the high suface energy of the air bubbles with the cell membranes. Sodium carboxymethyl cellulose (0.1%) is added to protect cells against mechanical damage and pluronic F-68 reduces foaming and so prevents cells from becoming stuck to the vessel walls above the medium level.

8.2.3 pH

Most cell lines require a pH of 7.4. Hybridoma and some tumour cells can survive at a pH as low as 6.8 but growth is impaired. The pH of the culture is influenced by the buffering components of the medium, the amount of headspace (in a closed system), and the concentration of glucose. Medium is normally buffered using a CO_2-bicarbonate system which is analogous to the physiological system. Phosphates in the medium also aid buffering and, for some cell lines, HEPES ([N-2-hydroxyethyl piperazine-N] ethansulphonic acid) can be added to provide further buffering.

The headspace in a closed, batch culture vessel should contain air supplemented with 5% (v/v) CO_2. Gas exchange occurs at the medium-air interface, and is further facilitated by the continuous stirring of the medium. However, as the cells grow and generate CO_2 there comes a point when gas exchange can no longer occur and the pH of the medium becomes acidic. A continuous flow of medium, as in perfusion cultures, prevents extremes of pH. For very large scale cultures, a pH probe inserted into the culture vessel can be connected to reservoirs of sodium bicarbonate or dilute sodium hydroxide which can be programmed to release enough alkali into the medium to maintain the required pH.

Such a pH control circuit is illustrated in Figure 8.1.

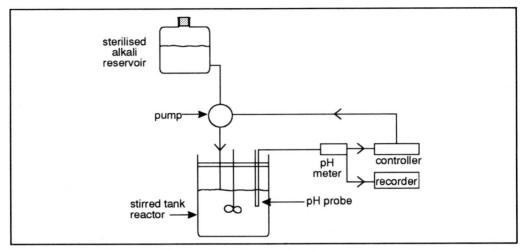

Figure 8.1 Schematic representation of a pH control circuit for a large scale stirred tank reactor. If the pH of the culture falls below a set value, the controller switches on the pump which allows alkali to enter the reactor. When the pH reaches the required value, the pump is switched off.

8.2.4 Oxygen

probability of oxygen depletion

Oxygen is very insoluble in medium (approximately 7.6 μg ml^{-1}) and cells typically use it at a rate of 5-7 μg min^{-1} 10^6 cells. Thus a culture of 2×10^6 cells ml^{-1} would very rapidly run out of oxygen. Clearly it is necessary to supply oxygen throughout the culture period. Surface aeration, sparging, medium perfusion or increasing the partial pressure of O_2 in the headspace can all help to maintain adequate oxygen levels in the system. Surface aeration can be achieved by passing gas through a nozzle into the headspace above the surface of the medium.

sparging and foam formation

Sparging is the bubbling of gas through the medium. This is a very effective way of oxygenating medium (the principle is the same as that used in fish tanks), but the froth that is produced can cause membrane damage and leave cells stranded high above the medium level. Antifoaming agents such as pluronic F-68 and lowering gassing rates help prevent cell damage.

With large scale vessels, it is usual to monitor dissolved oxygen concentrations and to control the sparging rates in order to maintain the oxygen concentration within set limits. A typical circuit is shown in Figure 8.2.

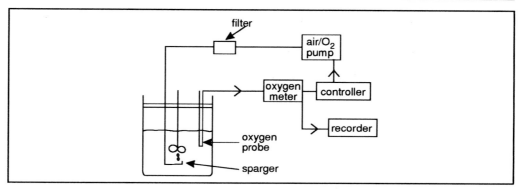

Figure 8.2 Schematic representation of a dissolved oxygen control circuit. If the dissolved oxygen falls below a set value, the air pump is switched on. When the dissolved oxygen reaches the required level the air pump is reduced to a pre-set minimum level.

∏ In the legend of Figure 8.2, we indicated that the air pump is switched to a minimum level when sufficient oxygen is dissolved in the medium. Why is it not switched off completely?

The reason is that if the air pump was switched off, medium (and cells) would enter the sparger and that may lead to fouling of the air line. By maintaining a small air flow, liquid is prevented from entering and fouling the sparger.

Perfusing pre-oxygenated medium through the culture is the most reliable way of maintaining adequate oxygen levels throughout the vessel. This approach cannot, however, be used in non-fed batch systems (see below).

∏ How would you determine what growth phase a culture is in when it is not possible to sample the cells?

A clue could be gained by measuring the oxygen or CO_2 in the inflow and outflow of the culture vessel. Such methods are, however, sometimes difficult to use. An alternative is to monitor the redox potential of the culture medium.

8.2.5 Redox potential

The redox potential of a culture is important. This potential is affected by the proportion of oxidising and reducing agents in the medium. The redox value falls during the logarithmic growth phase of the culture, and is at its lowest level at around 24 hours before the cells reach the stationary phase (see Chapter 5). Regularly monitoring the redox potential of the culture medium (with a probe) gives a good indication of the growth phase of the culture.

8.3 Scale up of anchorage-dependent cells

Cells which normally grow in suspension, such as human lymphocytes, are easier to scale up from a 1 litre culture through to a 1000 litre system than anchorage-dependent cells. Some anchorage-dependent cells can be adapted to grow in suspension, but human diploid cells are totally anchorage-dependent and must be grown in monolayer cultures which involves specific problems in scaling up.

Monolayer cultures:

- are often cumbersome and expensive to scale up;

- require more space than suspension cultures;

- need to be treated with trypsin to remove cells from the substratum before cells can be counted to assess growth;

- present some logistical problems in maintaining homogenous pH, nutrient and oxygen levels.

Nevertheless, as we will see later, many regulatory authorities stipulate that biopharmaceuticals for human use must be produced only in diploid cells, so it is essential to be able to grow these cells in large enough quantities.

8.4 Culture vessels

In this section we will describe a variety of culture vessels. It would be sensible to make yourself a summary table of this information as it will enable you to compare these vessels directly. We suggest you use the following format:

vessel type	typical surface area/volume of medium	comments
Roux bottle	175 cm^2 (100-150 ml)	cell yields 2 x 10^7 (diploids)
		1 x 10^8 (heteroploids)
Roller bottle	750-1500 cm^2 (200-500ml)	1-2 x 10^8 (diploids)
Roller bottle: spiral film	8500 cm^2 (_ _ _ _)	good with heteroploids or with diploid cells

8.4.1 Scale-up I - Roller bottles:

problems of scale-up using Roux bottles

Roux bottles or plastic flasks have only one surface for cell attachment so there is a low surface area to volume ratio. The largest Roux bottle or flask provides a surface area of 175 cm^2 and needs 100-150 ml of medium to grow 2 x 10^7 diploid cells or 1 x 10^8 heteroploid cells. If 1 x 10^{10} cells are required, one would need one hundred flasks and a 100 fold increase in incubator capacity. It would also involve a hundred manipulations for changing media, measuring cell viability or harvesting.

slow rates of rotations used with roller bottles

Roller bottles provide a surface area of 750-1500 cm^2 for cell attachment and growth and need only 20% of the cell sheet to be covered by medium at any one time (see Figure 8.3). Typically, using 200-500 ml of medium yields 1-2 x 10^8 diploid cells. The bottles are placed horizontally on rollers and are rotated at a rate of 10-20 rh^{-1} (rph) while cells are attaching and then at 6-8 rh^{-1} (rph). This system provides a much larger surface area than the culture flask and also ensures a much higher surface area to volume ratio. Gas exchange is also made easier because the cells are alternately aerated and fed with medium every few minutes.

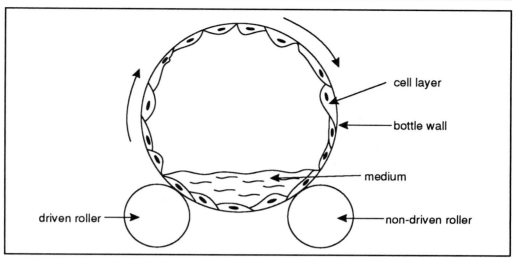

Figure 8.3 Transverse section through a roller bottle.

8.4.2 Scale up II: Increasing the surface area within the roller bottle

∏ As you can see from Figure 8.3 there is a great deal of wasted space in the roller bottle. How would you increase the surface area available for cell attachment without increasing the size of the bottle?

There are several ways of achieving this. Here we will describe two methods:

- the spiral film method;

- the glass tube method.

Spiral film

Figure 8.4 shows a roller bottle filled with a cartridge of spirally wound film of plastic, which increases the surface area available for cell attachment and growth to 8500 cm^2. Yields are increased by 5 - 10 times over the simple roller bottle. The bottle is rolled only while the cells are attaching, so there needs to be enough medium to cover the cells when the culture is stationary. As there is no headspace for air exchange, oxygen has to be provided by sparging or frequent changes with pre-oxygenated medium. Though this system works well with heteroploid cells results are often poor with diploid cells lines.

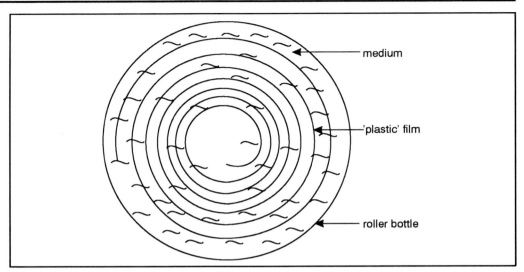

Figure 8.4 Transverse section through a Roller Bottle with spiral film to increase surface area.

Glass tubes

The principle is similar to that above, except that the surface area within the roller bottle is greatly increased by packing the bottle with parallel clusters of small glass tubes separated from each other by silicone spacer rings. Medium is perfused through the bottle which is rotated alternately through 360° clockwise and 360° anticlockwise to help maintain a homogenous environment (see Figure 8.5).

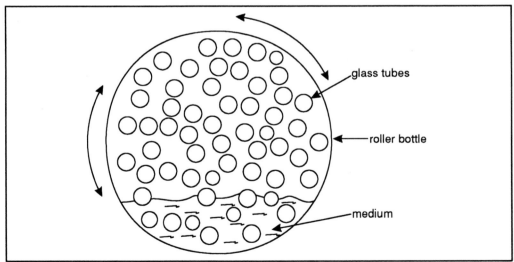

Figure 8.5 Transverse section through a Roller Bottle packed with glass tubes to increase surface area.

8.4.3 Scale-up III: Beyond the roller bottle

Multitray units

In multitray units, usually ten flat chambers, each with a surface area of 600 cm^2 are fixed together and interconnected at the corners by two vertical tubes (see Figure 8.6a). The tubes have apertures which allow the medium to flow between the compartments only when the unit is laid on its end. When laid flat each compartment is isolated from the others although the apertures still allow the transfer of gas throughout the system.

system uses
large volumes
of media

This 'cell factory' provides a total surface area of around 6000 cm^2 and has the advantage of not being very different from conventional flask cultures. One disadvantage, however, is that this system utilises a very large volume of medium (up to 12 litres).

Cells can be harvested by placing the unit on its end and draining off the medium, washing with buffer and trypsinising in much the same way as is used for 175 cm^2 flasks. The disadvantage of not being able to completely remove all the cells from the trays can be turned to advantage by adding fresh medium to the remaining cells which serve as inoculum for the next culture. This type of unit has been used to produce some interferons.

Hollow fibre culture

A vessel similar to the roller bottle is filled with bundles of hollow fibres which provide a surface for growth similar to that of the spiral film and glass tube roller bottle system. Once the fibres are packed in the bottle, medium is pumped in and perfuses through the fibre walls which are porous enough to allow the passage of macro-molecules of molecular weights between 10 000 and 1 000 000 Daltons. The fibres are made of acrylic polymer and are typically of about 350 μm in diameter with 70-75 μm thick walls. It is difficult to calculate the exact surface area but the surface area to volume ratio is very high: up to 30 cm^2 ml^{-1} supporting up to 10^8 cells ml^{-1}. An example of a hollow fibre culture system is illustrated in Figure 8.6b.

Many advances have been made in the design and production of fibres and this type of bioreactor is becoming increasingly popular especialy when cultures are used to produce specific excreted molecular products such as antibodies. The hollow fibre bioreactor design enables the product to be continuously removed in the outflow of the bioreactor. Further details of hollow fibre, and the related membrane, bioreactors are given in section 8.6.

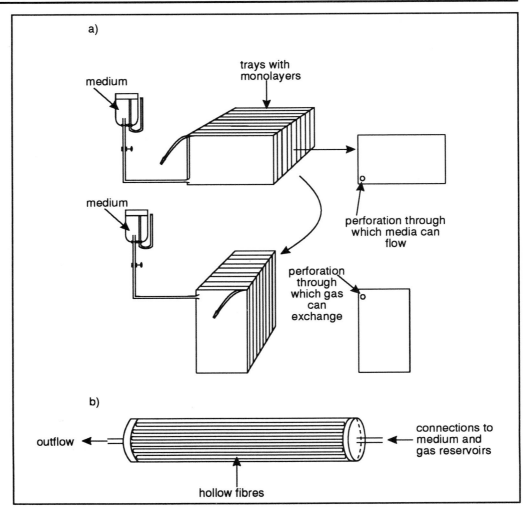

Figure 8.6 a) Multitray unit for the large scale culture of anchorage-dependent cells. The unit is filled with medium, then turned on its side to isolate medium in the chambers. b) Hollow fibre culture.

Plastic bags

use of gas
permeable
materials

Plastic bags for culturing cells are, in fact, made of fluoro-ethylene propylene copolymer teflon (FEP Teflon®, Du Pont) which is biologically inert but very gas permeable. A typical bag measuring 5 x 30 cm can be filled with medium to a depth of 2-10 mm, with cells attached to both surfaces. If this bag is placed directly in an incubator, the gas permeable walls will allow sufficient oxygen to diffuse through to sustain the culture throughout its life. Homogeneity of the medium can be maintained by rotating or rocking the bags very gently whilst in the incubator. Cells can be harvested by trypsinisation or by bending and stretching the bags.

Plastic film propagators

Here the same material, FEP Teflon®, is used but instead of bags, the plastic film comes in the form of a long tube wrapped round a reel, much like a camera film. Medium is pumped through the tube and recirculated through a reservoir or discarded if necessary. Gas exchange occurs through the walls of the tubing as with the plastic bags. The tubing can be up to 10 metres long, giving a surface area of up to 25 000 cm^2.

All the scale up systems described so far can be placed inside conventional incubators to maintain an even temperature. The next step up means more cumbersome reactor systems which need independent heating, filtration and gas exchange facilities.

8.4.4 Scale-up IV: Fermenter (bioreactor) systems

fermenters and bioreactors

Strictly speaking, vessels used to grow animal cells should not be called fermenters. Fermentation is an anaerobic process and the term fermenter should really be used to describe vessels in which cells are grown under anaerobic conditions. The term fermenter has, however, found fairly widespread use in describing vessels in which cellular material is grown. An alternative to using the term fermenter is to describe such vessels a bioreactors.

In bioreactor (fermenter) systems, anchorage-dependent cells still need to be grown attached to a substratum. Usually the substatum is kept in suspension. We can identify three basic types of bioreactors used to cultivate such cells on the basis of the nature of the substratum used. These are:

- glass bead reactors;

- stacked plate reactors;

- microcarrier systems.

In the following sections we will describe the general features of these systems. A more process technology-based discussion of these reactors covering aspects such as oxygen supply, substrate diffusion and product removal is given in BIOTOL technology texts ('Bioprocess Technology: Modelling and Transport Phenomena'; 'Operational Modes of Bioreactors' and 'Bioreactor Design and Product Yield').

Glass bead reactors

Glass beads of 3-5 mm diameter are packed tightly in a column through which medium is continuously perfused by means of a peristaltic pump. The cells grow on the surface of the glass beads. These are packed sufficiently tightly so as not to slide around and damage cells but with enough spaces to allow medium to flow freely through at a rate which is not likely to damage cells by setting up shear forces.

When the culture is confluent, the medium can be drained off, the beads washed with buffer and trypsinised.

∏ What disadvantage can you see with this system?

difficulties of sampling glucose utilisation rates

Cells cannot be sampled for monitoring growth or viability until the end of the culture. Neither can some of the cells be harvested at intervals during the culture. In this situation, growth can be monitored by measuring 'Glucose Utilisation Rates' since, when the cells are growing logarithmically, there is a very close correlation between glucose utilisation and cell numbers.

We can relate the rate of glucose (substrates) utilisation to the amount of biomass present by the equation

$$\frac{d\,[S]}{dt} = Q[X]$$

in which:

[S] = glucose (substrate) concentration

t = time

Q = specific rate of glucose (substrate) consumption = rate of substrate use per unit of biomass

[X] = biomass concentration

By measuring the rate of glucose utilisation (or respiration rate) we can determine [X] providing we know Q.

The relationship holds as long as Q is constant and substrate is not rate limiting.

use of substrate for growth and maintenance

Substrate is utilised to provide energy and cellular material for growth. It also provides energy for maintenance functions. For example, cells need to expend energy to maintain an osmotic balance and they use energy to repair damaged structures. These latter two processes, although essential, do not lead, in themselves, to any increase in biomass or cell numbers.

∏ Consider cells in the log phase; they use the glucose for maintenance and for cell growth. What do cells in the stationary phase use glucose for?

You should have predicted that they would only use glucose for maintenance purposes.

∏ What will happen to the value of Q as cells pass from log phase into stationary phase?

You should have anticipated that the cells would use less glucose. Thus we might anticipate that Q would fall in value.

Thus:

in the log phase the glucose is used for growth and maintenance (Q is high);

in the stationary phase the glucose is just used for maintenance (Q is low).

Whilst cells are in a balanced growth phase (such as, log phase) we might anticipate that Q (rate of glucose used per unit of biomass) remains constant. In the stationary phase Q will also be constant, but at a lower value.

Thus during the log phase, the rate of substrate consumption would follow the relationship $\frac{d\,[S]}{dt} = Q\,[X]$.

In stationary phase, the rate of glucose consumption would follow the relationship $\dfrac{d\,[S]}{dt} = Q'\,[X]$ where Q' is a new specific rate of substrate consumption characteristic of non-growing cells.

SAQ 8.1	1) In an experiment in which cells were cultivated in a glass bead reactor, medium containing $10\ g\ l^{-1}$ glucose as the main energy source, was pumped into the vessel at a rate of $100\ ml\ h^{-1}$. The glucose in the outflow was monitored and the following results were obtained.

time (h)	glucose concentration in the outflow (g l^{-1})
2	8.2
5	8.0
10	7.3
20	5.7
30	2.8
40	1.4
50	1.4
60	6.4
70	6.6

Interpret this data in terms of the growth of the animal cells. (Assume that the culture had not become contaminated and that the animal cells in the culture had remained viable).

2) A similar experiment to that described in 1) was carried out except in this case the medium contained $1\ g\ l^{-1}$ glucose. The following results were obtained.

time (h)	glucose concentration in the outflow (g l^{-1})
5	0.1
10	0.1
20	0.1
30	0.08
40	0.08
50	0.08
60	0.08
70	0.08

Are the cells in this culture in the logarithmic growth phase?

The growth yield of a culture (Y) relates the amount of biomass produced to the amount of substrate consumed. Mathematically it can expressed as;

$$Y = \frac{d\,[X]}{d\,[S]}$$

where d[X] is the change in biomass concentration and d[S] is the change in substrate concentration.

Y falls in
stationary
phase
During the log phase, Y usually remains constant. But once the cells enter the stationary phase, the value of Y will continue to fall since glucose continues to be used with no increase in biomass.

Stacked plate reactors

In this type of reactor, circular glass or steel plates are fitted 5-7 mm apart onto a central shaft. The shaft can be kept stationary while medium is moved across the plates by means of a pump, or alternatively the shaft can be rotated on a horizontal or vertical axis. The stacked plate reactor (see Figure 8.7) provides a surface area of up to 2×10^5 cm^2, but the medium volume to surface area ratio is very high (1 ml of medium to every 1.2 cm^2).

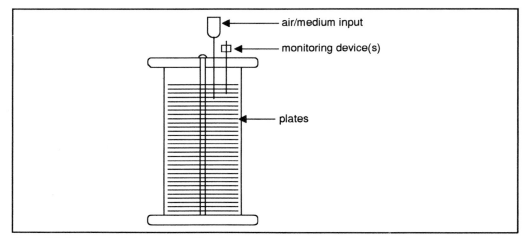

Figure 8.7 The stacked plate reactor. The plates may be rotated or medium pumped around the vessel in order to bring fresh medium into contact with the cells.

Microcarrier systems

many materials
used as
microcarriers
In microcarrier systems the cells are grown as monolayers on the surface of tiny spheres or cylinders which have a diameter of 100-200 μm, providing a surface area of 4.5-6.0 x 10^3 cm^2 per gram of spheres. A wide range of commercially produced microcarriers are available. Most are spherical in form though cylindrical forms are also available. DEAE sephadex, collagen, polyacrylamide, polystyrene and gelatine have all been used to make the microcarriers.

replenishment
of media
Microcarriers, with cells attached to their surfaces, can be maintained as a suspension culture by constantly stirring to keep the carriers in suspension and to maintain homogeneous pH and oxygen levels. As the cells grow, the spheres become heavier and begin to settle out if the stirring rate is not increased. The medium becomes acidic and depleted of nutrients very rapidly. Replacement of culture medium is easy in this type of culture: the stirrers are turned off and the spheres allowed to settle for 2-3 minutes, after which period the medium can then be decanted off and replaced as required.

separation of cells by differential setting out, filtration or solubilisation of carriers

Harvesting the cells from these microcarriers is also very simple. Once the medium is drained off, the microcarriers can be washed and trypsinised. The carriers are stirred rapidly for 20-30 minutes and the detached cells and microcarriers can be separated by allowing them to settle out. If all of the cells are to be harvested the mixture can be poured through a funnel with an appropriate sized filter so that the cells pass through leaving the microcarriers behind. Collagen coating on spheres can simply be dissolved by adding collagenase, and gelatine beads can be solubilised with either trypsin or ethylenediamine tetra acetic acid (EDTA).

Measuring the rate of growth or confluency of the culture is also very easy. A sample of the suspension can be siphoned off at any time for counting or morphological examination. Virus titres or growth factor concentrations can also be measured in this way at intervals during the culture. Virus titres are, of course, important if we are using the cell culture to raise viruses to produce vaccines or for research. In some cases virus titres are also important in checking for contamination.

purfusion of microcarrier systems

The microcarrier concentration can be increased to some extent in order to increase cell yield, but a higher cell density means that the medium is depleted of nutrients and oxygen more quickly and the pH falls sharply. For large scale microculture, the medium needs to be perfused through the system. Filters allow medium to be removed without losing the carriers. Stirring speeds have to be kept low to prevent frothing which can leave microcarriers stranded high above the medium level. Providing adequate amounts of oxygen can also be difficult. Oxygen can be delivered to the system in one of three ways:

- directing a stream of gas (95% air, 5% (v/v) CO_2) over the surface of the culture for a few minutes when taking samples or changing the medium;

- filling the headspace with oxygen-containing gas when sampling. Continuous stirring of the suspension ensures that gas exchange can occur at the medium surface;

- perfusing with medium, which has been oxygenated in a chamber outside the culture vessel.

8.5 Suspension culture

Scale up is much easier for cells which can be grown in suspension. Some anchorage-dependent cell lines can be selected or adapted to grow in suspension, but not the human diploid cells which are so valuable for producing biopharmaceuticals.

Cell lines derived from haemopoietic tissues (for example H9, Hut 78, Raji) and hybridoma lines normally grow in suspension either as single cells or in small clumps of cells derived from single cells. Suspension cultures can be scaled up from a flask to a deep tank culture.

Here we will describe two types of deep tank reactors. First we will examine conventional stirred tank reactors and then we will describe looped reactors. These latter types use a variety of devices to induce a cyclical flow of culture in the vessel.

Scale up can, however, be achieved using two broadly different strategies. The most obvious is to increase the volume of the vessel. Alternatively, scale up can be achieved by increasing the biomass concentration in the reactor. We will examine these in latter sections. We will also describe how we can adapt batch systems so that they can be used in continuous flow operations.

8.5.1 Stirred tank bioreactors

Typically, laboratory-scale cultures are produced in spinner flasks. They derive their name from the action of a small bar magnet driven by an external magnetic stirrer. In principle, such vessels can be used for volumes from 20 ml to 20 litres. In practice, however, such vessels are difficult to handle once the volume exceeds about 5 litres. For larger volumes we may, therefore, use a stirred tank bioreactor of the type illustrated in Figure 8.8.

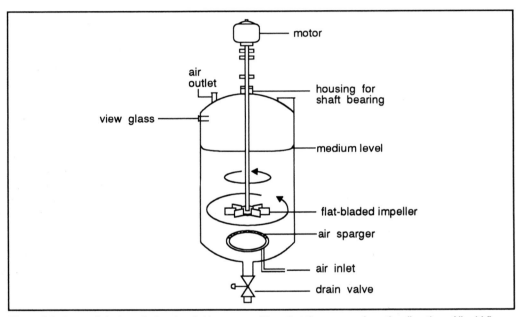

Figure 8.8 General design of stirred tank bioreactors. Note that the arrows show the direction of liquid flow.

The impellers used in these vessels can be of various designs. The most common designs are the disc type, turbine type and marine type shown in Figure 8.9.

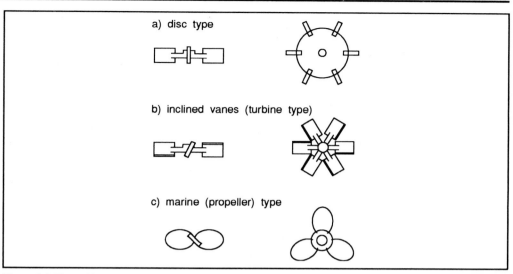

a) disc type

b) inclined vanes (turbine type)

c) marine (propeller) type

Figure 8.9 Typical impellors used in stirred tank bioreactors. a) disc, b) turbine, c) propeller.

∏ See if you can identify an advantage of using a stirred tank bioreactor of the type illustrated in Figure 8.8 over using a spinner culture.

advantages of stirred tank reactors compared to spinner cultures

The main advantage is that the stirred tank bioreactor allows for better control of the mixing of the culture. Properly designed impellers are more efficient mixers than bar magnets! The more efficient control of mixing obtained in stirred tank bioreactors has important consequences for maintaining homogeneous culture conditions. There is a lower probability of setting up pH gradients or areas of low oxygen tension in such vessels.

Stirred tank reactors and looped reactors are available up to a working volume of 8000 litres although for the gas lift types, the commercially available vessels are usually produced up to a volume of 1000 litres.

8.5.2 Changes which occur on scale up

The scale up from the small spinner cultures used in laboratories to the bioreactors used on an industrial scale (for example 1000 litres) involves a number of important changes. Here we will deal with the following five issues:

- the materials used to construct the vessels;

- the method of temperature control;

- the supply of oxygen;

- the requirement to be able to sterilise the vessel and its connections *in situ*.

Materials used for large vessels

Generally, the large culture vessels are made of stainless steel rather than of glass or plastic.

∏   What are the dangers in using stainless steel instead of glass or plastic?

The main danger is that these metal vessels will allow leaching of metal ions into the medium. These metal ions may be toxic to the animal cells being cultivated. Thus special steels that have been prepared/treated to prevent these toxic ions from leaching into the medium have to be used.

Temperature control in large vessels

∏   How can the temperature of the culture be maintained at the appropriate level in larger vessels?

In contrast to smaller vessels which can be placed in an incubator, larger vessels have to be either equipped with a water jacket or with internal heating/cooling coils.

∏   See if you can list some advantages and disadvantages of using water jackets rather than internal heating/cooling coils.

limited area for
heat exchange

The advantage of using an external water jacket is that it does not occupy any of the working volume of the reactor. The main disadvantage of these devices is that the area available for heat exchange is limited. This is not a major problem with smaller vessels but it can become a problem with very large vessels. Consider for example a cylindrical shaped reactor. If the volume of the reactor is doubled, the surface area of the reactor is not doubled. (You might like to prove this for yourself. Remember that the volume of a cylinder $= \pi r^2 h$ and if we assume that the area for contact with a water jacket is limited to its vertical walls, its surface area $= 2\pi rh$. By substituting in values for r and h, you will see that the surface:volume ratio declines as the value of r is increased).

Thus, we might anticipate that the available surface for heat transfer could become rate limiting in very large vessels. This could, therefore, mean that there is localised overheating or overcooling within the vessel. We will not elaborate on this further here but the problems of heat transfer and the determination of the temperature differential needed between water jacket and medium to maintain a particular reactor temperature are dealt with in the BIOTOL text 'Bioprocess Technology: Modelling and Transport Phenomena'.

The alternative to using an external water jacket, is to use internal heat exchangers. These are usually tubular coils or plates through which water can be pumped. These types of heat exchangers are illustrated in Figure 8.10.

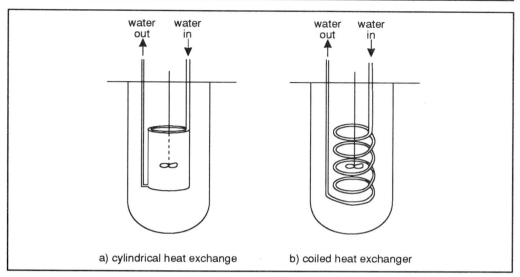

Figure 8.10 Diagrammatic representation of internal heat exchangers.

Π What are the main advantages and disadvantages of using internal heat exchangers of the types illustrated in Figure 8.10?

advantages and disadvantages of internal heat exchangers

The main advantages of these types of heat exchangers are that they can be designed to give a very large surface area for heat exchange. Thus the temperature difference between the water circulated through the heat exchanger and the medium can be kept small. Thus there is less likelihood of getting localised overheating/overcooling within the medium. The main disadvantages of these devices stem from the fact that they reduce the working volume of the reactor and that they increase the difficulty of harvesting cells from the reactor. Cells tend to adhere to these devices. Reactors containing internal heat exchangers are more difficult to clean than those fitted with external heat exchangers.

Oxygen supply

oxygen limitation in large vessels

One of the main factors limiting the scale up of cultures is oxygen availability. With small cultures, the surface area to volume ratio is large. Thus, although oxygen transfer rates across the air:medium interphase are low (171 µg cm^{-2} h^{-1}), the large surface area to volume ratio enables sufficient oxygen to diffuse into the medium. In larger volume reactors, there is a lower surface area:volume ratio and thus if oxygen is allowed merely to diffuse in through the air:medium surface oxygen in the medium will become limited.

There are two fundamentally different ways of increasing oxygen transfer into the liquid. These are by increasing the stirring rates and by sparging. Thus, in scaling up from a small spinner culture to a stirred tank reactor we have to consider whether to increase the stirring rate or to use sparging or to use a combination of both.

Π What problems do you foresee with increasing the stirring rate?

We hoped that you would realise that this would set up increased shear forces which may damage the cells.

☐ What problems do you foresee by increasing the sparging rate?

Again, high gas velocities through spargers set up localised shear forces which may damage the cells. It may also cause foaming.

In practice stirrer rates up to about 300 rpm may be used and gas flow rates of about 5 ml min^{-1} l^{-1} should not be exceeded.

oxygen limitation less likely with animal cells then with bacteria

Since animal cells do not grow to high cell densities and their rates of metabolism per g biomass are relatively low compared with bacteria, oxygen limitation is not usually a major problem. Low stirrer rates and low sparging rates are usually sufficient to maintain oxygen at an appropriate level. With bacterial cultures, the high cell densities (often 10^{10} cells ml^{-1}) and high rates of metabolism present greater difficulties in maintaining appropriate oxygen levels. Fortunately these cells are more tolerant to shear forces and they can withstand high stirrer and sparging rates.

SAQ 8.2

The surface area of a culture grown in a spinner culture vessel is 100 cm^2. The rate of oxygen consumption by the culture is calculated to reach 5000 µg h^{-1} towards the end of the log phase of growth. Oxygen is supplied in the form of air pumped into the head space of the vessel and its rate of transfer across the air:medium interface has been determined to be 170 µg cm^{-2} h^{-1}.

It is proposed to use a scaled up version of this spinner culture. The culture volume is to be increased 10 fold. The vessel that is to be used has a cross sectional area of 200 cm^2.

What problems do you foresee in supplying oxygen to this culture and how would you solve these problems?

Sterilisation of large vessels

sterilisation *in situ*

Large vessels cannot be placed in autoclaves, so sterilisation has to be done *in situ*. This is usually achieved by connecting the vessel to a steam generator which injects super-heated steam into the vessel. It is not only essential that the reactor itself is sterilised but also the feeder and outlet pipes must also be sterilised. Probes (for example for pH, temperature) also require sterilisation. The sterilisation of pipes, taps and valves is also achieved using super-heated steam. This, of course, needs careful reactor design, since if any part of the system is not properly sterilised, this may act as a potential source of contamination.

The design of the sterilisation system is largely an engineering problem. The engineer has to ensure that the steam reaches all parts of the system and that all parts of the system are maintained at sufficiently high temperatures to ensure sterility.

8.5.3 Looped bioreactors

A variety of looped bioreactors are shown in Figure 8.11. These reactors have many of the features of stirred tank reactors (for example control of oxygen concentrations, pH, sterilisation *in situ* etc) but differ in that they are designed to induce a cyclical flow of culture around the vessel.

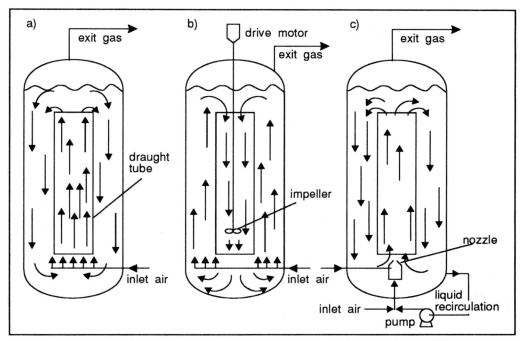

Figure 8.11 Examples of looped reactors. Cyclical flow may be induced by a) gas-flow (air-lift), b) an impeller or c) liquid pumping (jet-flow).

air-lift, impeller and jet types of looped reactors

In the gas-lift (air-lift) type, air enters the vessel via a sparger near the bottom. This creates small bubbles in the medium near the bottom of the draught tube. The medium:bubble mixture is of lower density than medium alone, so it begins to rise. The movement of medium in the impeller (propeller) type is, of course, driven by the action of the impeller whilst in the jet-flow type, the liquid:air mixture pumped into the vessel provides the momentum to keep the medium in the vessel well mixed.

Π   Of the devices shown in Figure 8.11, which do you think is most suitable for cultivating animals cells? (Justify your selection).

All three types of devices are efficient in supplying oxygen and all can be fitted with pH monitors and pH controllers. The main differences in the vessels relate to the shear forces set up in the medium. The main shear forces are, of course, set up close to the energy input into the system (that is, at the sparger in the gas-lift, at the impeller in the impeller type and at the nozzle in the jet-flow type). Generally bioreactors of the gas-lift type are regarded as being the most gentle and are often selected for cultivating animal cells. This does not mean, however, that the other types cannot be used but care has to be taken not to damage cells.

air-lift often preferred

8.6 Increasing cell densities and continuous flow cultivation

low cell densities achieved

The bioreactors described above all suffer from one serious problem. The vessels as we have described them yield fairly low cell densities. Typically, they would yield cell concentrations of up to $3\text{-}5 \times 10^6$ cell ml^{-1}. To achieve sufficient cell yields for many commercial operations we would, therefore, need to use extremely large vessels.

However, if the concentration of cells within the reactor could be increased, the reactor volume could be reduced.

∏   See if you can suggest ways in which the concentration of cells in such reactors could be increased.

The solution to this is to find ways of supplying fresh medium to the reactor while retaining the cells within the reactor.

We can represent this problem in the following way:

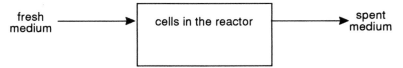

There are several ways in which this might be achieved. Perhaps the simplest way is to periodically stop aggitating the medium and allow the cells to settle. Then a portion of the spent medium could be withdrawn and replaced by fresh, pre-warmed medium.

∏   What problems can you foresee with this approach?

problems with
using a settling
out-decantation
approach

The main problems arise through nutrient and oxygen starvation during the settling out. Freely suspended animal cells take quite some time to settle and as they do so they pack together at the bottom of the vessel. Since the supply of oxygen to the vessel has been switched off, the oxygen in the small volume of medium surrounding the cells becomes rapidly consumed and cellular respiration is stopped. The cells, therefore, stop growing and may be further damaged by their inability to provide the energy needed for maintenance functions. Thus, not only is culture time lost during the settling out period, but also the viability of the cells may be dramatically reduced and the cells may take considerable time to recover.

A potential solution to this problem is to physically separate the cells into a separate compartment(s) and to use a medium perfusion technique. This can be achieved in the following ways:

*   by using membrane reactors;

*   by using hollow fibre cartridges;

*   by encapsulating the cells in a porous polymer.

Let us consider membrane reactors. The basic design of these is shown in Figure 8.12.

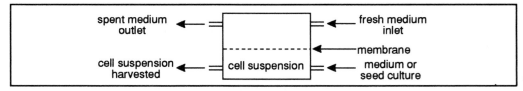

Figure 8.12 Diagrammatic representation of a membrane reactor.

In a membrane reactor, continuous cultivation is possible. Fresh medium is fed into one compartment and is allowed to diffuse across a membrane into the compartment containing the cells. The cells in this second compartment can be continually harvested.

It is important to have a large interphase between the medium and the cell suspension, so these reactors are usually designed in the form of multiple chambers as shown below:

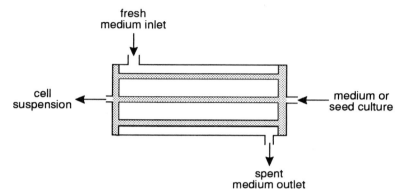

Instead of using flat bed membranes, the membranes can be arranged in the form of hollow fibres. In these cases, medium is pumped through the hollow fibres whilst the cells are allowed to grow outside of the fibres.

The advantages of these membrane reactors is that cells sensitive to shearing remain productive for a long time and can be grown to high densities. Yields of over 10^8 cells ml^{-1} can be achieved.

Π How is oxygen supplied to the cells in membrane reactors?

It has to be supplied in the perfusing medium. Thus the medium has to be aerated before being pumped into the reactor. It has to be pumped at a sufficient rate to ensure that the cells do not become depleted of oxygen.

Π At the beginning of this section, we suggested that by increasing the density of cells in the reactor, we could use a smaller reactor to achieve the same cell yield. Does this mean that we can use equipment of small volumes for the process?

Although we can use a smaller reactor, remember that 1 ml of medium will support only a limited number of cells (usually about 1×10^6 cells). Thus the medium reservoir has to be of the order of 100 times the culture volume. In effect, what we have done by using a membrane reactor is to replace a large volume reactor by a small volume reactor and a large medium reservoir.

Π What then is the real advantage of using this approach?

The real advantage stems from the high concentration of cells. It means that we have to harvest smaller volumes of culture and this facilitates downstream processing.

The commercial application of membrane reactors is, however, not straightforward. This is due to the unpredictability of the membranes. Membranes tend to clog quite often and have a limited life span varying between six months to two years. Furthermore, membranes are relatively expensive. Nevertheless, tremendous advances are being made in the preparation, composition and properties of membranes and you should anticipate that this technology will find progressively more extensive commercial use.

8.6.1 Continuous flow bioreactors

In the previous section, we indicated that membrane and hollow fibre reactors may be employed as continuous flow systems. Here we extend our discussion of continuous flow bioreactors to encompass the other types of bioreactors.

Consider the reaction vessel set up as a batch culture as shown below (we have omitted pH, O_2 and other monitoring and control devices for simplicity).

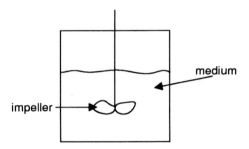

If we inoculate this culture, the cells begin to grow. Eventually they will become depleted of a particular nutrient (the growth-limiting nutrient) and the growth rate will slow down. Eventually it will stop. If we now begin to pump in fresh medium and at the same time remove culture at the same rate, the cells in the vessel will begin to grow again. The amount of growth is dependent upon the amount of limiting nutrients that is pumped in.

We can represent this situation in the following way:

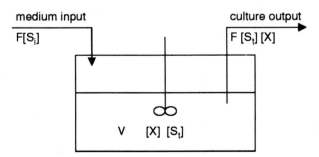

where F is the flow rate into and out of the vessel, V = volume of the culture; [X] = concentration of the biomass in the vessel; $[S_i]$ = substrate concentration in the input and $[S_t]$ = substrate concentration in the culture.

We can write a balanced equation for the biomass in the vessel in the following way:

rate of change of biomass in the culture	=	increase in biomass by growth	−	loss of biomass via the outflow

Thus, mathematically

$$\frac{d[X]}{dt} = V\mu[X] - F[X]$$

where μ = specific growth rate of the biomass (= rate of increase in biomass per unit of biomass).

If $\mu[X]$ is greater than $F[X]$ then the biomass concentration in the vessel will increase. This will continue until a steady state is reached in which the rate of biomass removal equals the rate at which it grows.

In this situation $\frac{d[X]}{dt} = 0$

and $V\mu[X] = F[X]$

If we divide through by V and [X] then

$$\mu = \frac{F}{V}$$

$\frac{F}{V}$ is known as the dilution rate and is given the symbol D.

$\mu = D$ Thus at a steady state $\mu = D$. In other words, the growth rate equals the dilution rate when a steady state is reached.

In principle, therefore, we can grow the cells at any growth rate we desire providing we do not exceed the maximum specific growth rate (μ_{max}) of the cells. We can simply control the growth rate by adjusting the dilution rate of the incoming medium.

A continuous culture in its steady-state can be considered as a self-regulating system. For instance, if the growth rate becomes smaller than the dilution rate ($\mu < D$), the biomass concentration temporarily decreases (a negative balance exists). As a result, the substrate concentration in the vessel increases. This, in turn, causes μ to increase again until a new balance is reached (ie $\mu = D$).

In practice, it is usually assumed that the steady state is reached when the biomass concentration has not changed over a period in which a total of five volume replacements have taken place since the new dilution rate was set.

The attraction of this system is that not only biomass concentrations and growth rates are maintained at a steady state but that the chemical/physical environment of the cells also remains in a steady state.

A more advanced mathematical treatment of continuous cultures is given in the BIOTOL text 'Bioreactor Design and Product Yield'.

With animal cell cultures dilution rates are usually low (D is usually given in units of day^{-1}) when compared with microbial cultures (D usually in h^{-1}). Nevertheless, continuous cultures are useful for many types of animal cells including LS, HeLa-cells and lymphoblastoid cells.

Continuous culture systems of the type described above are often referred to as chemostats or biostats reflecting the chemical and biological steady state achieved in these vessels.

SAQ 8.3

1) A stirred tank bioreactor has a working volume of 5 litres and is being used in a continuous flow format. A steady state biomass concentration of 1×10^6 cells ml^{-1} is reached using a medium input rate of 8 litres day^{-1}. What is the specific growth rate and what is the biomass yield day^{-1}.

2) The flow rate used in 1) was increased to 20 litres day^{-1}. After 2 days, the biomass concentration had fallen to 2×10^5 cells ml^{-1} and was still declining. Explain why this had occurred.

8.7 Selecting the appropriate bioreactor system

In the above sections, we have described a variety of formats of bioreactors that can be used to cultivate animal cells in suspension. We have learnt of some of the advantages of using each type. In this section, we will try to summarise these.

We have learnt that scale up can be achieved by either increasing the culture volume or by increasing the cell density per unit volume. The vessels can be used either in batch format or in continuous flow format.

In general, batch technology is derived from methods used to cultivate micro-organisms. It is associated with large vessels, low cell densities and high capital investment. We have learnt that stirred tank reactors suffer from problems that arise from the turbulence and shear forces generated at the tips of the impellers. These problems can be alleviated to some extent by using air-lift reactors. Nevertheless, all batch systems suffer from a potentially serious problem. Culture in batch format often means that the protein, antibody or growth factor one is trying to produce is retained in a deteriorating culture environment. Loss of biological activity may occur as a result of proteolytic degradation, inhibitors, oxidation and other processes. If antibiotics are included in the medium to reduce the chances of microbial infection, they have to be used at relatively high 'single dose' concentrations.

Similarly antifoaming agents have also to be used at relatively high concentration and may also be difficult to remove from the end product. This raises questions about the safety of the products to be used for therapeutic purposes.

In selecting a bioreactor we, therefore, have to consider a very wide set of issues. It is difficult to give a universally applicable check list to use in making such a decision but the major issues are:

- to make a comparison of the capital and operational costs that would be incurred using various reactor formats;

- to evaluate the relative ease/difficulty of isolating the product with the required specifications from cultures using the various reactors;

- to evaluate the relative reliabilities of the various systems that are available;

- to evaluate issues of safety to operatives and the public using the various reactor systems that are available;

- to consider the adaptability of the system to generate alternative products.

Under each of these headings, there are, of course, many sub-questions that need resolving. For example, in making a comparison of costing, we would need not only to consider the costs of the equipment, energy, feedstocks and labour, but we need to include calculations of product yield and quality as well.

By attempting the following SAQ, you will be able to demonstrate to yourself that you have understood the principle advantages and disadvantages of the various bioreactors we have described.

SAQ 8.4	Select from the list of bioreactors given below those which match the following statements. 1) Cell densities are usually low. 2) Products have a long residence times in cultures. 3) Require the treatment of large volumes during downstream processing. 4) May generate high shear forces which damage cells. 5) Require large media reservoirs. 6) Present a high probability of microbial contamination. 7) Enable growth of cells in a constant chemical environment, 8) Can be operated continually over a long period. **Bioreactors** Spinner cultures, batch stirred tank reactors, air-lift reactors, membrane/hollow fibre reactors, continuous flow stirred tank reactors.

8.8 Applications of large-scale animal cell cultures

In the previous sections, we have described some of the many types of bioreactors available for scaled up culture of animal cells. These bioreactors find extensive use in industry. Industrial uses of animal cells are mainly to produce vaccines and therapeutics. To give you some idea of the range of products, we have listed some examples in Table 8.1. Included in this table are data describing how many cells need to be cultivated to produce a single dose of vaccine or therapeutic.

If we take polio vaccine for example, 2×10^4 host cells are sufficient to produce a single dose of vaccine. This can be supplied from about 0.0001 litre of animal cell culture. In contrast urokinase, used as an antithrombolytic agent, requires 500 litres of animal cell culture to produce a single dose.

Product	Cells required/dose	Culture volume required/dose (l)
Vaccines		
Polio	2×10^4	0.0001
Rabies	4×10^6	0.005
HSV	2×10^7	0.03
FMDV	2×10^7	0.01
Therapeutics		
Interferon (anti-viral)	10^5/day	0.1
Monoclonal antibodies	10^{12}	100
Urokinase	10^{12}	500

Table 8.1 Products made by animal cell culture. Note that these are only examples, the exact dose size depends on the particular quality of the product. Data derived from Katinger and Bleim (1983), Production of enzymes and hormones by mammalian cell culture Adv Biotech Proc 261.

∏ Based on these values, would there be a larger requirement for industrial bioreactor capacity to produce polio vaccines or to produce urokinase?

Your intuition might be to suggest that urokinase requires a larger industrial commitment of bioreactor capacity. This is not necessarily so. Remember that a large proportion of the population is vaccinated against polio and only a small proportion of the population receives urokinase treatment. The point we are making is that the bioreactor capacity that needs to be committed to making a product depends upon three main factors:

• the yield/unit volume of bioreactor;

• the amount required per dose or course of treatment;

• the number of individuals who require the product.

Π Let us carry out a specific calculation. It is proposed to produce sufficient polio vaccine to vaccinate 5×10^6 individuals annum^{-1}. In the process to raise the vaccine, cells are cultivated for 4 days before harvesting. About 10 000 doses of vaccine can be produced per litre of culture. What size of reactor would be needed to produce the annual requirement.

If it takes 4 days to cultivate the host cells, then we could, in theory, produce $\dfrac{365}{4}$ cultures annum^{-1} = 91 cultures per annum.

Thus each culture would be required to produce sufficient vaccine for $\dfrac{5 \times 10^6}{91}$ individuals = approximately 6×10^4 individuals. As 1 litre of culture produces 10^4 doses of vaccine, we could, in theory, produce sufficient vaccine using $\dfrac{6 \times 10^4}{10^4}$ litres cultures = 6 litres. In practice, however, we need to use much larger culture volumes than this. Firstly, because we will rarely achieve 365 successful working days per annum. Secondly, some cultures may be lost through contamination and/or other production problems. Thirdly, we will need additional material to use for quality control and standardisation.

Π It is anticipated that 10^3 doses of urokinase will be required per annum. Again let us assume that the cells are cultivated for four days before harvesting. What size of reactor would be required to produce the annual requirement.

Using an analogous approach to our calculation for the polio vaccine, we again calculate 91 cultures could be produced per annum.

From the data given in Table 8.1, we require a total of 500×10^3 litres of culture to produce 10^3 doses. Thus the volume of bioreactor must be at least $\dfrac{500 \times 10^3}{91}$ litre = (approx) 6×10^3 litres.

In practice, we would again have to allow for down time and losses incurred.

The two examples we have used (polio and urokinase) are two extremes, but they have served to illustrate the type of calculation that need to be done.

Now attempt a similar calculation for yourself.

SAQ 8.5	A single dose of interferon can be obtained from 0.1 litre of culture. Assume that individuals treated with interferon are given single doses on each of ten days. You have a batch bioreactor with a working volume of 2×10^3 litres and you estimate that you can run 50 cycles of culture per annum. How many patients may be treated with your product?

The data given in Table 8.1 are, of course, dependent on how successfully the product is recovered from the culture. With very valuable products (especially those which are made with low yields and are required in high doses such as urokinase) the efficacy of downstream recovery is of vital importance.

extension of the range of products made in animal cells *in vitro*

Let us now turn to another aspect. The growth of recombinant DNA technology (see Chapter 10) is having two main impacts on the industrial cultivation of animal cells. First it is rapidly expanding the range of products made using animal cells cultured *in vitro*. Some examples are given in Table 8.2. It is also changing the formulation of media used to grow cells. In general, by genetically manipulating animal cells, it becomes possible to reduce the need for complex media supplements.

Group	Examples
Vaccines and related products	Hepatitis Bs Antigen (HBs Ag), HS viruses 1 and 2, Influenza vaccines, potential vaccines for Rabies, HIV, Lassa fever
Hormones	Human growth hormone, luteal hormone, erythropoietin, insulin, relaxin
Blood products	Immunoglobulins, blood clotting factors (factors VIII and IX)
Interferons and interleukin	β-IFN, Interleukin 2

Table 8.2 Recombinant products made by animal cells.

adoptive immunotheraphy and organ replacement therapy

Not only are animal cells used to make useful biochemical products, cells themselves may be used. For instance, large numbers of cells need to be produced for use in adoptive immunotherapy. In this form of therapy, cells are removed from a patient's body, grown in the presence of cytokines to boost cell numbers, and returned to the same patient. Organ replacement therapy also benefits from scaled up tissue culture. Cells from liver, skin, pancreas and adrenals have been cultivated in the laboratory in order to augment organs whose own capabilities are inadequate.

8.9 Regulatory issues

No products made from animal cells can be put to clinical use without appropriate market authorisation. Such a market authorisation is achieved by providing documentary evidence about the purity, consistency, potency, safety and benefit of the product. This is administered nationally by the appropriate regulatory authority. Increasingly there is a growing tendency for national groups to act within the framework of supra-national organisations. For example within the EC, extensive use is made of providing multiple market authorisation through a central organisation.

This text is not the place to deal with the issues of market authorisation in depth. This is done in the BIOTOL text 'Biotechnological Innovations in Health Care'. Nonetheless, you need to be made aware that the production and use of animal cell-derived products must be done within the guidelines of Good Manufacturing Practice (GMP). This includes regular checks of plants, laboratories and products and a clear set of quality control measures implemented. Within the EC, the procedures for carrying out quality control assays are specified by a series of guidelines.

We will only mention DNA and viruses here. Because of the potential for animal cells to harbour retroviruses which may transfer oncogenes (see Chapter 6), specific measures have to be taken to ensure viruses are removed from mammalian cell products. Nucleic acids which may contaminate these products must be demonstrated to be at concentrations below 10-100 pg/dose.

Summary and objectives

In this chapter, we have briefly examined the reasons why we may need to culture animal cells on a large scale. We divided our discussion of scale up strategies into the scale up of anchorage-dependent cells and the scale up of anchorage-independent cells. In the latter case, we explained that scale up might involve increases in culture volume and/or increase in cell density. We described a wide range of reactor configurations and discussed the advantages and disadvantages of the various types.

Towards the end of the chapter, we cited some specific examples of products made using animal cell culture and indicated the importance of recombinant DNA technology to the future development of animal cell cultivation and application. We concluded the chapter by briefly referring to the need to meet market authorisation requirements for animal cell-derived products.

Now that you have completed this chapter you should be able to:

- describe a wide range of strategies for scaling up anchorage-dependent and anchorage-independent cell cultures;

- list the advantages and disadvantages associated with various reactor types;

- calculate bioreactor volumes required to achieve desired product yields from supplied data;

- give examples of commercial products made using animal cell cultures;

- use suitable examples to explain how recombinant DNA technology is having an impact on the practical application of animal cell culture techniques.

Three-dimensional culture techniques

9.1 Introduction   186

9.2 Organ culture   186

9.3 Reconstruction of three-dimensional structure   188

9.4 Uses of embryonic tissue   190

9.5 Embryo culture   191

9.6 Applications of IVF and embryo transfer   198

Summary and objectives   203

Three-dimensional culture techniques

9.1 Introduction

So far, the techniques described in this book have involved mainly the growth of dispersed cells either in suspension or in monolayer culture. Although these techniques have many advantages (discussed in Chapter 1), the loss of structural integrity of the tissue may significantly alter the functional properties of these cells. Many general aspects of cells can be studied using this model, but the integrated function of the whole organism requires more complex models. As a result, culture methods have been developed in which the arrangement of cells resembles that *in vivo*.

∏   Before we look at these methodologies in more detail, see if you can remember what factors regulate the differentiated status of cells *in vivo*.

soluble factors, extracellular matrix and direct cell interactions

These were covered in Chapter 4 (section 4.4.2) where we mentioned three types of mediators of cell differentiation; soluble factors (such as hormones and paracrine factors); changes in the composition of extracellular matrix; and direct cell interactions (via gap junctions or surface information exchange). For the most part, these features are missing in dispersed cell culture and in this chapter we discuss more complex culture techniques in which much of the physiological environment is either retained or reconstructed.

In the first approach, the three-dimensional structure of the tissue is retained by explantation and growth of organs or parts of organs so that architecture is preserved and outgrowth of isolated cells is minimised.

∏   We defined this technique in Chapter 1. See if you can remember what it is called without looking back.

organ culture

The three-dimensional culture of tissue that is handled so that it retains some of the normal histological features of that tissue is called organ culture.

An alternative approach is not to try to preserve the *in vivo* architecture of an organ, but to actually reconstruct the three-dimensional structure *in vitro*.

In this chapter we will discuss both methods before describing the use of embryonic tissues and embryo culture. This will enable us to consider 'reproductive' technology and to examine the ethics of manipulating embryos.

9.2 Organ culture

∏   What are the disadvantages of organ culture compared to cell culture?

Generally, organ cultures are more difficult to prepare than cell cultures. Once a cell line has been isolated and cloned, preparing fresh cultures simply requires inoculation of fresh medium. Organ cultures, however, cannot be propagated. Thus each experiment requires fresh tissue from the donor source. The production of such explants has to be done carefully in order to retain the structural integrity of this tissue.

organs cannot be propagated

Experiments using organ culture generally involve a large degree of experimental variation between replicates, so that reproducibility is less easily achieved than in cell culture. Organ culture is essentially a technique for studying the behaviour of integrated tissues rather than isolated cells and structural integrity is the main reason for adopting organ culture as an *in vitro* technique in preference to cell culture. As such, organ culture lies somewhere between cell culture on the one hand - and experiments using whole animals on the other - both in terms of the types of results which can be obtained and the complexity of the experimental model. As with all experimental models, caution should be exercised in the interpretation of results. Sometimes the systemic factors are important, for example a drug may be metabolised *in vivo* but not *in vitro*, thus giving qualitatively and quantitatively different results in the two situations. There is also time limitation on organ culture. Since these cultures cannot be maintained indefinitely any effects which appear only after several months will not be observed using this model.

experimental variation and finite life of organ cultures

When cells are cultured as a solid mass of tissue, a number of problems arise which are not an issue with cell culture. In particular, gaseous diffusion and the exchange of nutrients become limiting. To overcome these problems, a number of different techniques have been developed. Each of these have their limitations. We will discuss these later.

9.2.1 Historical development of organ culture

You may recall from Chapter 1, that one of the earliest tissue culture experiments was performed in 1885, when Roux maintained the medullary plate of a chick embryo in warm saline for a few days. In 1897, Loeb was the first to culture fragments of adult rabbit liver, kidney, thyroid and ovary on small plasma clots inside a test tube. He found they retained their normal histological structure for three days.

It is now known that the best results are obtained by growing the tissue on a solid support within a fluid medium. The organ culture is kept at the interface between liquid and gaseous phases to facilitate gas exchange while retaining access to nutrients. There has been a steady development of new and improved methods to provide these conditions.

solid support fluid interface

The first substrate used was a clotted plasma substrate. In 1929, Fell and Robinson introduced the 'watchglass technique', by which organ rudiments or organs were grown on the surface of a clot consisting of chick plasma and chick embryo extract contained in a watchglass (see Figure 9.1).

clotted plasma watchglass technique

This became a standard technique for morphogenetic studies of embryonic organ rudiments and the method was later modified to investigate the action of hormones, vitamins and carcinogens in adult mammalian tissues. The main disadvantage of this technique is that explants often liquify the clot and sink into a pool of liquified medium. For this reason the technique was superseded by the agar clot technique, which uses agar instead of clotted plasma, in the 1940s.

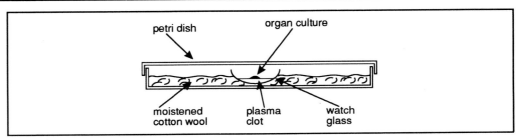

Figure 9.1 Diagram to show the watchglass technique of organ culture.

The next major development occurred in 1954, when lens paper (used for cleaning microscope lenses) was used as a floating raft on which cultures were grown. It was possible to study cell-cell interactions by growing different types of tissue on either side of the lens paper.

The use of rafts floating on a fluid medium had a number of disadvantages. The rafts often sank and the tissues were frequently immersed in the medium. This difficulty was overcome by the 'Grid technique' which used metal grids made of wire gauze (and later more rigid expanded metal) instead of lens paper. Tissues are cultured directly on the grid or on strips of lens paper which are deposited on the grid. The method gives tissues a firm support which prevents them from being submerged by the medium. The grids with their explants are placed in a culture chamber filled with medium up to the level of the grid (see Figure 9.2). The method is widely used for embryonic and adult tissues and succeeds in preserving the histological structure of the tissues.

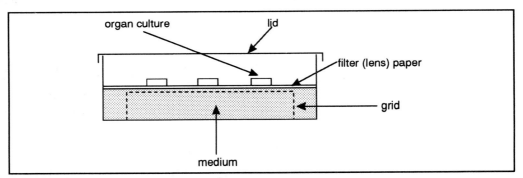

Figure 9.2 Diagram to show the Grid technique of organ culture.

9.3 Reconstruction of three-dimensional structure

In organ culture, organs or parts of organs are grown in culture so that they retain the histological features of the tissue *in vivo*. There are other culture methods which start with dispersed populations of cells but encourage the arrangement of these cells into organ-like structures. In other words rather than preserving the architecture of an organ, the three-dimensional structure is actually reconstructed *in vitro*. In these systems important physiological factors may be quantified thus overcoming some of the disadvantages of organ culture.

We can sub-divide these types of culture into two basic types:

- histotypic cultures;

- organotypic cultures.

9.3.1 Histotypic culture

histotypic culture

Culture systems using one characterised cell line propagated at high density in the presence of appropriate extracellular matrix and soluble factors are called histotypic. For example, vascular endothelial cells will form capillary tubules in the presence of appropriate soluble factors when grown in a collagen matrix. Homologous cell interactions occur because the cells are cultured at high density. An alternative approach is the use of cellulose sponges coated with extracellular matrix components such as collagen. Cells may penetrate the sponge and form glandular structures.

9.3.2 Organotypic culture

organotypic cultures

Histotypic cultures use one cell type and, therefore, exclude the study of heterologous cell interactions which are so important in initiating and promoting differentiation. When cells of different lineages are recombined to create a tissue-like structure the culture is called organotypic.

co-culture

At its simplest such a culture system may involve the co-culture of two cell types. For example co-culture of epithelial and fibroblast cell clones derived from the mammary gland allows the epithelial cells to differentiate functionally as demonstrated by their ability to produce milk proteins in the correct hormonal environment. This functional differentiation is preceded by the formation of characteristic structures such as three-dimensional cords in which fibroblast cells organise themselves into bundles which are then enveloped by epithelial cells.

organoids

As well as applications in research, the generation of organ-like structures from dispersed cell lines may one day be used in the treatment of human disease. In 1989 the first working 'organoid' was reported - an artificial liver surgically implanted into the peritoneal cavity of a rat. The organoid was made from a combination of cells, growth factors, collagen and Gore-tex fibres. Such organoids could potentially replace diseased organs or deliver genetically altered cells into a patient's body.

9.3.3 Categorising tissue culture

You may have noticed that the different categories of tissue culture methods defined in this book are actually overlapping. This reflects the range of experimental approaches to the *in vitro* cultivation of animal cells. Although it is convenient to use such a classification it is important to recognise that there are overlaps between these methodologies. Attempt the following SAQs as they will test your understanding of these terms.

SAQ 9.1	Label the different types of tissue culture in the list below according to whether you would describe them as cell culture, organ culture, histotypic or organotypic culture.

1) Monolayer culture of a pure cell strain on plastic.

2) Monolayer culture of a pure cell strain on collagen.

3) 3-D culture using a pure cell strain grown at high density.

4) 3-D culture using more than one cell strain.

5) 3-D culture using more than one cell strain and 3-D matrices.

6) 3-D cultivation of tissue so as to retain its architecture.

SAQ 9.2	Now assign the following tissue cultures to an appropriate type (eg cell culture, organ culture etc).

1) Growth of a pure cell strain on a 'feeder' layer of a different cell type.

2) Recombination of different tissue types (from tissues, not from dispersed cells).

3) Monolayer culture of primary cell culture (ie culture derived from an explant).

4) Whole embryo culture.

9.4 Uses of embryonic tissue

We have made a number of references to the *in vitro* cultivation of embryonic and foetal tissue.

∏ List the different uses of embryonic and foetal tissue mentioned in this text.

embryonic tissue to provide explants for cell culture

In Chapter 1 we discussed the cell cultures which can be derived from embryonic tissue. In general, cell lines from embryonic tissue survive and grow better than those from adult tissues. Cultures of rodent embryo cells routinely give rise to continuous cell lines whereas Hayflick's experiments using human foetal lung fibroblasts showed that human foetal cells have only a limited capacity to grow in culture, since the cultures invariably degenerate. Nevertheless, human embryonic cells in culture have had a number of uses including the generation of vaccines for mass vaccinations against polio.

A more recent development has been the removal of embryonic stem cells (ES-cells) from the embryo during the blastocyst stage of development. These cells can be grown in culture for many generations and are of particular interest because they can be manipulated in culture and then re-introduced into embryos.

embryonic tissue as a source of organ culture

In addition, much organ culture has been performed using embryonic tissue. Organ rudiments can develop *in vitro* in a similar manner to *in vivo*. The earliest tissue culture experiments were performed using the nervous tissue of chick embryos (Roux's experiments) and many of the studies of early development were carried out using organ culture.

However, the use of embryos is much broader than being merely sources of cells for cell and organ culture. Intact embryos are of tremendous importance and in the next section, we will examine the production and cultivation of embryos.

9.5 Embryo culture

These days, whole embryos can be grown *in vitro* and a high proportion of these embryos will develop well in culture. The term embryo culture is used to describe the *in vitro* development or maintenance of isolated embryos. This term excludes cell culture and organ culture, and refers simply to the cultivation of intact embryos *in vitro*.

definition of an embryo

Before we go on to the details of embryo culture, let us pause to consider exactly what is meant by the term embryo. This is actually not that easy, since the word embryo has been used to mean different things by different people. Clearly the word refers to a stage of development before birth. The broadest definition of the word embryo may be applied to structures that exist starting with the fertilised ovum and ending with birth. Strictly speaking, however, this is incorrect. An embryo is a spatially defined entity that can develop directly into a foetus. In the foetus, the main structures of the body are formed, so the word embryo implies that all the main structures of the body have not yet been formed. Thus the word embryo is often used to denote the developmental stage of an organism starting with the fertilised ovum (the zygote) and ending with the foetus. Some people say that the term embryo is even more limited than this, and refers to the stage between the blastocyst (a spherical mass consisting of a single layer of cells from which the inner cell mass projects into a cavity) and the foetal stages. The exact details of these arguments are beyond the scope of this text, and it will suffice to know that in this chapter we are referring to the *in vitro* cultivation of an animal in the early stages of development, prior to birth.

In Figure 9.3, we remind you of the important stages in the development of an embryo. This figure is, of course, a great simplification of the processes of embryo development but is sufficient for our purpose here.

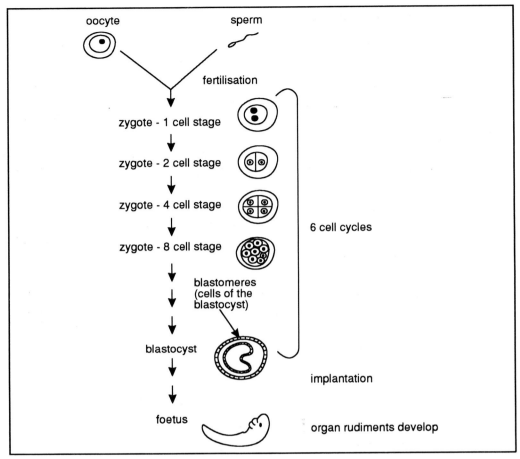

Figure 9.3 Stages in the development of a mammalian embryo.

9.5.1 Sources of embryos

There are two main sources of embryos for embryo culture:

- by dissection;

- by *in vitro* fertilisation (IVF).

Dissection of embryos

One source of embryos is pregnant animals - for example if a pregnant mouse is killed, the uterus can be removed aseptically and the embryos can be retrieved from the uterus. One cell stage embryos can be isolated from the ampullary region of oviducts from mated female mice on the date of vaginal plug detection. The main problem is the removal of embryos in an undamaged state. This requires careful dissection and the embryos can then be washed and cultured under silicone oil at 37°C.

In vitro fertilisation to obtain embryos

In invertebrates vast numbers of gametes and embryos are available for experimental studies due to the natural occurrence of external fertilisation.

The other source of embryos is to create them *in vitro* by fertilising oocytes from the female using sperm from the male, in other words by *in vitro* fertilisation (IVF). The technique of IVF was developed primarily in mice and hamsters to study the process of fertilisation and early development.

In mammals, minute numbers of eggs and embryos are produced in each cycle by females. However, females can generally be induced to superovulate using gonadotrophins.

mammals can be induced to superovulate

Fertilised ova can be grown to the blastocyst stage *in vitro*, and the embryos apparently undergo normal development since re-implanted embryos can go on to form healthy animals. The technique of IVF has been applied to a large number of species during the past 25 years with varying degrees of success.

The ability to induce superovulation depends upon knowledge of the hormonal control of the oestrous cycle. The most common way is to use treatments with progesterone and injections of gonadotrophin. The exact treatment is dependent upon the species of animals being used. A detailed description of the induction of superovulation in a variety of animals, especially domestic animals, is given in the BIOTOL text 'Biotechnological Innovations in Animal Productivity'.

Here we will give an outline scheme for the production of embryos *in vitro*.

9.5.2 Production of embryos *in vitro*

An overview of the production of embryos *in vitro* is given in Figure 9.4. Use this figure to follow the description given in the text.

Induction of superovulation

In many mammals only a single follicle or a few follicles in the ovary develop to produce mature oocytes. Yet in most mammalian ovaries there are many follicles each capable of producing oocytes. Three types of follicles can be distinguished. Some are large (10 mm in diameter), some are medium in size (2-10 mm), while the remainder are small (less than 2 mm). Often 50-80 follicles of diameter 2-10 mm develop during the normal oestrous cycle. However, all but a few degenerate. If the animal is injected with gonadotrophic hormones such as follicle stimulating hormone (FSH), the development of several follicles may be maintained. In this way, many oocytes may be released during ovulation. These superovulation procedures, however, still suffer from highly variable yields.

maintenance of follicular development using FSH

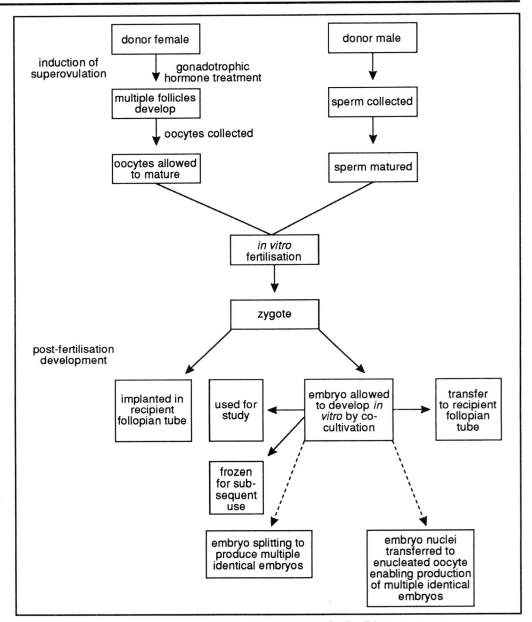

Figure 9.4 Overview of the *in vitro* production of embryos (see text for details).

One of the alternatives to superovulation is the *in vitro* maturation and fertilisation of oocytes. The method that yields best results is to use transvaginal ultrasound guided puncture of follicles. This uses ultrasound to guide a needle via the vagina into the follicles to extract the oocytes. The method is, however, generally only applicable to larger animals such as cattle. The oocytes are then matured *in vitro*.

∏ Write down the major advantage that *in vitro* maturation might have over superovulation.

This procedure has the attraction that immature oocytes can be collected repeatedly from the same animal irrespective of the day of the oestrous cycle.

It is becoming common practice to use a combination of these two strategies. Superovulation is induced by gonadotrophic hormone treatment and the oocytes are normally collected 2-4 hours before ovulation is expected. They are then allowed to mature *in vitro* for 1-6 hours before insemination.

∏ Why are the oocytes collected before ovulation?

It is much easier to locate the follicles (and therefore the oocytes) than it is to locate oocytes released after ovulation.

Fertilisation *in vitro*

maturation of sperm

Fertilisation is achieved *in vitro* by the addition of sperm. Often the sperm has to be pre-treated to make it competent in fertilisation. For example, ejaculated bovine sperm is not able to fertilise the oocytes *in vitro*. It first needs to be treated with heparin or other substances like caffeine and lysophosphatidyl-choline to become mature. Little is known about the processes which occur during this maturation but much effort is being directed to elucidating them. It is anticipated that this research will lead to better ways of preparing competent sperm.

Post fertilisation development

This can be done in one of two ways. One way is to transfer the fertilised oocytes (zygotes) into the fallopian tube of a recipient animal. Alternatively, the oocytes may be allowed to develop *in vitro* by, for example, co-culture with tubal cells. The overall success of this method is fairly low. Usually, with cattle, about 25% of the zygotes develop.

Asexual multiplication of embryo by splitting or nuclear transfer

Each embryo, regardless of the method of production, contains a unique combination of the genotypes of its parents. Identical embryos can be produced by embryo splitting. In this technique, early stage embryos are first halved and then quartered. Each quarter is still capable of developing into a viable young. This process of embryo splitting is illustrated in Figure 9.5.

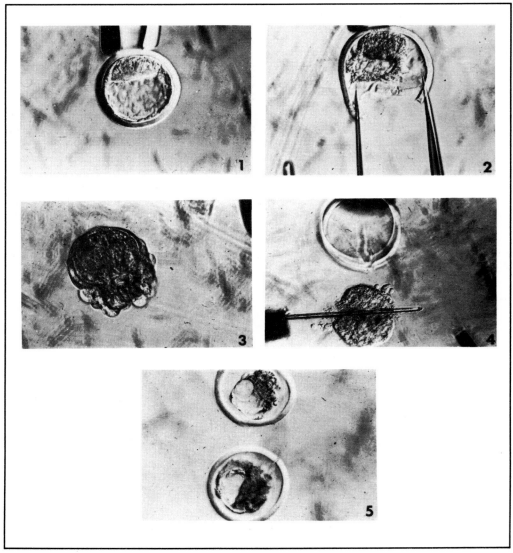

Figure 9.5 Embryo-splitting. The embryonic cell mass is first 'sucked out' of its protective membrane (1, 2 or 3) then split (4) to form the new embryonic cell masses (5).

Efforts to divide an embryo into eight parts have not succeeded. It appears that the number of cells is then too small to support the development of complete new embryos.

An alternative approach to producing a number of identical embryos is to use nuclear transplantation. Since each nucleus of an embryo, irrespective of the number of cells, has the same chromosomes present, each has the potential to support complete embryo development. If these nuclei are transplanted into enucleated oocytes, then multiple, identical embryos should be produced. Although still in its infancy, this technique is likely to be widely used with domestic animals.

Storage of embryos

storage in
liquid nitrogen

Obviously, producing embryos *in vitro* is expensive and a technically demanding process. It is, however, possible to freeze embryos in liquid nitrogen and, after thawing them, implant them many months or years after fertilisation. Since 1983, it has become possible to freeze and store human embryos in this way. However, freezing excess embryos is prohibited in some states and countries.

Re-examine Figure 9.4. You can see that the zygotes produced by *in vitro* fertilisation may be used in a number of different ways. If they are allowed to develop into embryos *in vitro*, then they may be used to study the early stages of embryo development. The embryos may also be transferred to a suitable host to give rise to foetuses and hence to mature offspring. Alternatively, these embryos may be split to produce multiple embryos for use in research or to produce multiple identical offspring. In the latter case the 'split' embryos are transferred to suitable hosts. This technique has been successfully used to produce identical farm animals but ethical and legal considerations (see below) exclude its application to humans.

| SAQ 9.3 | When zygotes or embryos are transferred to recipient fallopian tubes, what points must be taken into consideration in selecting suitable recipients? |

Evaluation of embryo quality.

A variety of criteria may be used to evaluate the quality of embryos that are to be used for transfer or cryopreservation.

This criteria may include items from the following list:

* has the embryo achieved the right levels of development it should have reached at the time of harvesting?

* are the blastomeres (cells of the blastocyst) uniform?

* are the membranes distinct and intact?

* are vesicles present?

* are there signs of degeneration and dissolution?

On this basis embryos may be classified as excellent, good, fair or poor. In other words, these criteria are qualitative and rather imprecise.

invasive
techniques

Ratings, based on physiological parameters, have also been used. These include dye-exclusion tests and measurement of key enzymes. These tests are, however, invasive.

assay of
metabolites

Much effort is being put into developing non-invasive techniques. Recently, a non-invasive technique has been developed based on the changing metabolic requirements of embryos. In mice, it has been shown that pyruvate is an essential nutrient during the first cleavage division. Glucose is unable to support development until the 8-cell stage. Thus, using microfluorescence-enzymatic methods to determine small changes in the concentrations of pyruvate and glucose in microdrops of media used to cultivate embryos provides a basis for evaluating their quality. It appears that pyruvate requirements of human embryos are similar to those of mouse embryos, so this test may also be used to evaluate the quality of human embryos.

9.6 Applications of IVF and embryo transfer

In this section we will briefly describe the practical applications of IVF and embryo transfer under four headings:

- applications in embryology;

- applications in transgenics;

- applications in animal breeding;

- applications in human reproductive technology.

We should, however, point out that the details of many of these applications are provided elsewhere in the BIOTOL series of texts. We will cite the appropriate texts in the relevant sections given below so that you are able to follow these up in greater depth if you so wish.

9.6.1 Applications in embryology

applications in embryology

The development of the mammalian embryo in the protected environment afforded by the female reproductive tract presents a natural obstacle to research. Thus the use of *in vitro* techniques for the study of early mammalian development is easier and more informative than attempting to study the events in the natural environment. Manipulation of embryos is an important tool in the experimental embryology of mammals and embryos have been used to provide information about normal development, growth and differentiation and the influence of extrinsic factors such as toxic agents in teratological studies. Much has been learnt about the mechanisms which control normal mammalian development using the animal embryo as an experimental model and these animal models have provided information which had been relevant to understanding human development.

9.6.2 Applications in transgenics

incorporation of DNA via male pronucleus

In 1980, results of experiments were published in which DNA was micro-injected into the male pronucleus of fertilised mouse eggs. The injected eggs were re-implanted into pseudopregnant females to develop, and it was shown that the micro-injected DNA had become incorporated into the chromosomes of the resulting newborns. Although no expression of this incorporated DNA was detected, subsequent studies reported high levels of expression of micro-injected fusion genes using similar techniques. Integrated genes were shown to be transmitted through the germ line and progeny were also shown to express the genes. Animals which contain integrated foreign DNA

applications in transgenics

are generally called transgenic and the foreign genes are conventionally referred to as transgenes. Subsequently the technology has advanced, and we have already mentioned the use of embryonic stem cells in generating transgenic animals in Chapter 1. There are countless applications of transgenic methodology including research into genetic diseases and improvements in productivity of animal livestock.

Genetic engineering techniques applied to animal cells are dealt with in Chapter 10. The applications in transgenic animal production are also dealt with in the BIOTOL text 'Strategies for Engineering Organisms'.

9.6.3 Application in animal breeding

The technique of IVF followed by embryo transfer is transforming animal breeding since it facilitates sex selection (for example allowing the selection of only females for dairy herds) and intercontinental shipping of embryos in place of adult animals. The transfer of bovine embryos from prize cows, each worth thousands of pounds, has become a significant industry.

embryo sexing   The sexing of embryos is usually based on the detection of sex-specific antigens on the surface of the cells. Alternatively sex-specific DNA probes may be used to detect sex-specific nucleotide sequences. Although this technique is reliable, a single blastomere still needs to be removed from embryos for testing purposes.

The approach of using embryos for experimental objectives is full of controversy and it is important to be aware of the relevant legislation before performing any experiments. For example, recent changes in the UK mean that use of embryos and foetuses beyond 50% gestation or incubation comes under the Animal Experiments (Scientific Procedures) Act of 1986. You should be acquainted with the legal status of embryo culture in your country if you plan to use this technique.

9.6.4 Applications in human reproductive technology

'test tube babies'   In recent years there has been a renewal of interest in IVF procedures stimulated in part by the successful use of this technique for the treatment of human infertility. The first child conceived with the help of the IVF technique was born in July 1978, and since then the IVF technique has led to the birth of almost 1000 babies (popularly called 'test tube babies') world-wide.

ovulation and the maturation of oocytes   In human ovaries at birth there are about 2 million cells which can mature into potential egg cells (ova). The complete maturation of these oocytes requires hormonal signals and generally one such maturation will occur in a monthly cycle. All other potential oocytes remain in a resting state or degenerate. The original steps in the maturation process occur in follicles within the ova. Follicles become fluid-filled, expand to about 2 cm in size and bulge out of the surface of the ovary. When the follicle ruptures, the maturing oocyte is propelled towards the fallopian tube. At this stage the cell is called a secondary oocyte. The maturation of this secondary oocyte into an ovum with the correct amount of genetic material is completed when the spermatozoan penetrates the membrane of the cell. When the maternal and paternal chromosomal materials mix, the zygote is produced. The first cell division produces 2 blastomeres, which may separate to produce identical twins. Within a few days after fertilisation, the ball of dividing cells has travelled through the fallopian tube and enters the uterine cavity.

In some individuals, however, the events between ovulation and implantation of the embryo in the uterine cavity are restricted or prevented. For example, if the fallopian tubes are blocked, the oocyte cannot enter the tube and thus fertilisation cannot be achieved.

These, and similar problems, may be overcome using IVF and *in vitro* embryo culture.

We will examine the applications of these techniques to solving human infertility problems. It will in part help to re-enforce the general scheme we described in section 9.5.

IVF in humans
laproscopy

An early step in IVF is the removal of the oocytes from their follicles by laproscopy. The oocytes are normally aspirated 2-4 hours before ovulation is expected and allowed to mature *in vitro* for 1-6 hours before insemination. More than 85% of pre-ovulatory oocytes from patients with tubal disorders are successfully fertilised *in vitro*. The embryos are allowed to develop *in vitro* to the blastocyst stage and then re-introduced into the female for implantation, where about 20-30% of the replaced embryos result in pregnancy.

follicle
stimulants

In order to increase the efficiency of treatment, the IVF programme generally includes ovarian stimulation to initiate maturation of more than one oocyte (superovulation). The woman is treated with a combination of clomiphene and human menopausal gonadotrophin (HMG) which act as follicle stimulants. Doctors use these oocytes to grow several human embryos *in vitro* and usually only some of these are used for implantations. Upto 10 embryos have been replaced on a single occasion. The embryos not implanted can be discarded, frozen down for future pregnancies, or used for research.

evaluation of
viability

A number of techniques have been used to assess human embryo viability. These are either invasive (resulting in destruction of the embryo or parts of the embryo) or non-invasive (for example observations on the rate of cell division). Since chromosomal defects have been observed in otherwise apparently normal 8 cell embryos, many would argue that more invasive tests should be performed to avoid implanting embryos with genetic defects.

research using
human material

Experimental research on early human embryos is an emotive issue and the development of the IVF technique has raised many ethical problems. A public debate has centred on the question of whether research using human embryos should be permitted at all, and many people consider that it is unethical to create a human embryo for any purpose other than enabling a child to be born. Others argue that the possible benefits far outweigh any such considerations and contend that human embryos provide unique models for research into:

• the diagnosis of infertility;

• improved treatments of infertility by IVF;

• diagnosis of genetic disease;

• the causes and treatment of congenital malformations.

The ethical and legal issues surrounding human embryo research are subjects of major controversy. We do not intend to take a moral stand here. We must, however, alert you to the fact that work in this area is increasingly becoming subject to legislation and guidelines. Perhaps in this area of scientific activity more than any other emotions and fears run high and there is great disparity between the positions adopted by different countries and states. These reflect religious and cultural differences. What is acceptable in one state/country may be unacceptable elsewhere.

human embryo
research and
legislation

In the UK, for example, the legal and ethical issues raised by IVF and the possible research which could be performed on unused human embryos were made the subject of a committee of inquiry set up by the Department of Health and Social Security and chaired by Dame Mary Warnock. The Warnock Committee report (1984) makes a number of specific recommendations, one of which is that research on human embryos should be permitted upto the 14th day after fertilisation. However, the issue of whether such studies are acceptable has not yet been resolved and, in the UK, several attempts have been made to completely prohibit embryo research. The delay in the resolution of this problem reflects the difficulty of generating consensus views and the rapid development of these technologies. Some would argue that research should be allowed

consensus is
difficult to
achieve

to continue so that the full capabilities of these techniques are assessed before legislation is formulated. In this way, we would be aware of the potential benefits (and potential dangers) before the imposition of regulations. Others would argue that this research is totally unethical and should be banned immediately. Of course, this would remove the moral and practical problems that might arise from these activities. At the same time, it would prevent the realisation of the benefits that might be accrued.

Individuals working in this area, will, of course, come to their own conclusions. Nevertheless, it is important for researchers to check that all proposed experiments fall within the appropriate laws of their country. For any experiments using human material, there may be legislation concerning consent from hospital ethical committees, attending physicians or surgeons, and the 'parents' of the embryo.

The future direction of human embryo research

surrogate
parenthood

The development of IVF procedures has facilitated further treatments of infertility, including the possibility of donating eggs or embryos to women who are unable to produce their own oocytes because of primary ovarian failure or women who are known to carry genetic defects. This allows the male to participate genetically and the female to carry a normal pregnancy. The implantation of a fertilised egg into the uterus of another woman, is called surrogate parenthood. In 1987, a 48-year old grandmother successfully carried 3 foetuses from her own daughter's oocytes.

In addition, the removal of the early steps of reproduction from the setting of the woman's body has opened up new possibilities for parents which include

- selecting the timing of birth. Foetal chromosomal disorders increase 2-3 fold when women are over 40 compared with those just a few years younger. Now that methods of freezing and thawing embryos are available, the time between oocyte retrieval and implantation may be extended allowing parents to choose the time of pregnancy;

- determining the sex and number of children;

- carrying our pre-implantation diagnosis and gene replacement;

- selection for specific traits.

Again, these new options are accompanied by a myriad of ethical, legal and social concerns.

| SAQ 9.4 | Which of the following techniques for evaluating embryos may be regarded as invasive and which are non-invasive? |

1) Measuring viability by a dye-exclusion test.

2) Evaluation of membrane status.

3) Evaluating the presence of vesicles.

4) Sexing, using a sex-specific DNA probe.

5) Sexing, using a fluorescently labelled sex-specific antibody.

6) Determining metabolic uptake of nutrients by embryos as a measure of their quality.

Summary and objectives

In this chapter, we have examined a range of culture types in which the spatial relationships of cells are important. These types of cultures include organ culture, histotypic cultures, organotypic culture and embryo cultures. We explained that in organ culture the spatial relationships and functional activity of tissue components are preserved *in vitro*. Thus organ culture is a more physiological model than cell culture, and yet has some of the advantages of *in vitro* work, namely that results are obtained more quickly than *in vivo*, that results can be readily quantitated, and that results are not complicated by the presence of systemic factors. Overall it can be said that organ culture is useful in conjunction with some confirmatory experiments using whole animals. We moved on from organ culture to histotypic and organotypic culture, where *in vivo* spacial relationships were reconstructed rather than preserved. Finally we discussed embryo culture, using embryos which had been removed from the pregnant female or created *in vitro* using IVF. Animal embryos have been used as an experimental model for understanding early development, and are also becoming economically important in animal husbandry. In humans, some types of infertility can be overcome by removing oocytes, fertilising them outside the womb, and replacing the embryo in the womb for the rest of the pregnancy. The use of superfluous embryos in research has brought public attention to focus on the IVF procedure and associated technologies, but detailed moral arguments are beyond the scope of this text.

Now that you have completed this chapter, you should be able to:

- describe the approaches used to study cell interactions and the integrated function of organs *in vitro*;

- define organ culture, organotypic culture, histotypic culture and embryo culture;

- specify the methods which have been developed to optimise organ culture;

- indicate applications, including possible future applications, of various three-dimensional culture techniques;

- describe the stages used in the generation of embryo cultures;

- list techniques/criteria used to evaluate the quality of an embryo;

- discuss the uses of embryo culture including *in vitro* fertilisation and embryo transfer.

Genetic engineering of animal cells

10.1 Introduction   206

10.2 Why choose animal cells to produce specific proteins?   207

10.3 What kinds of products are made using animal cells cultured *in vitro*?   208

10.4 Gene manipulation   209

10.5 Introduction of the recombinant DNA   227

10.6 The host system   231

10.7 The characterisation of transfected cells   232

10.8 Medical applications of genetic engineering   234

Summary and objectives   237

Genetic engineering of animal cells

10.1 Introduction

genetic
engineering

Genetic engineering is a relatively new methodology which was developed in the mid-1970s. At this time a succession of new methods evolved which facilitated the formation of new combinations of genes by allowing us to isolate DNA fragments, change them in defined ways, and re-introduce them into the same organism or to introduce them into a different organism. This is the essence of what has appropriately been called genetic engineering.

gene
manipulation

One aspect of genetic engineering is the *in vitro* manipulation of genes by the investigator. Gene manipulation has had to be defined legally as a result of government legislation to control it. The UK definition is 'the formation of new combinations of heritable material by the insertion of nucleic acid molecules produced by whatever means outside the cell, into any virus, bacterial plasmid, or other vector system, so as to allow their incorporation into a host organism in which they do not naturally occur, but in which they are capable of continued propagation'. Other national definitions may differ in the exact words used, but are very similar in sentiment.

recombinant
DNA
technology

The techniques needed for gene manipulation are called recombinant DNA techniques. Thus genetic engineering, recombinant DNA technology, and gene manipulation are related terms, and each term refers to a slightly different aspect of the overall picture, although the words are sometimes used interchangeably.

applications of
genetic
engineering

The genetic engineering of animal cells in culture has many possible applications. In general, genetic material is introduced into cells so that they will synthesize a protein which they would not normally synthesize. The objective may be to analyse the effects of a specific gene within a cell, or to produce a protein which has been processed correctly so that the protein can be further studied, or to correct a genetic defect in a human patient. The ability to introduce cloned genes into cells is a very important and powerful method, playing a critical role in many areas of biotechnology and medicine.

In this chapter, we will first consider why, in certain circumstances, it is more appropriate to use animal cells than bacterial cells to manufacture particular gene products. This will naturally lead us to consider the range of products which are most successfully manufactured using animal cells. We will then discuss the techniques used to manipulate genes. This will include the isolation of genes, their insertion into suitable vectors and the control of their expression. You should realise this is a large and rapidly moving area of research. Here we will cover the main concepts and techniques. If you wish to follow up this discussion to gain an in depth appreciation of genetic engineering in general, we recommend the BIOTOL texts 'Techniques for Engineering Genes' and 'Strategies for Engineering Organisms'.

10.2 Why choose animal cells to produce specific proteins?

Ⅱ Name one reason why you would choose to express a protein in an animal cell rather than a bacterial cell?

animal cells vs bacterial cells

Although bacteria are in many ways easier to work with to produce proteins on a large scale, the main problem with this system is that the genetically engineered products of bacterial cells are not necessarily identical to the products of the same gene when expressed in eukaryotic cells.

post-translational modifications

Let us look at this point in a bit more detail. One of the main obstacles is that bacterial cells process proteins in a different way from eukaryotic cells. To become mature proteins, polypeptides must fold into their native conformations and any subunits must properly combine. Proteolytic cleavage may be necessary and there is a number of covalent modifications which may need to take place for example, glycosylation (additional of carbohydrate residues), acetylation, hydroxylation, methylation, nucleotidylation, phosphorylation, ADP ribosylation, and formation of disulphide bonds. Most of these modifications do not play a role in bacterial proteins, and so the relevant enzymes are generally missing in bacteria. Many of these post-translational modifications do not even occur in yeast or other lower eukaryotes and animal cells may have to be used to produce authentic proteins. In addition, foreign proteins over-expressed in bacteria often form clumps of denatured proteins and many foreign proteins are broken down by bacteria. These problems do not generally occur in animal cells, and furthermore animal cells can be made to secrete their products into the cell culture medium, making purification easier.

There are, of course, also many disadvantages of working with animal cells. They are more fragile, slower to grow, have more complex nutrient requirements, and great care is needed to avoid contamination of cultures.

SAQ 10.1

Use a sheet of paper to produce a table listing the advantages and disadvantages of bacterial cell culture and mammalian cell culture in the large scale production of proteins. Use the following format.

	Advantages	Disadvantages
Bacterial cell culture		
Mammalian cell culture		

Before we move on, it is worth mentioning that even if animal cells are used there may be problems in getting the correctly processed protein product. This is because there are differences in the ability to glycosylate proteins (in other words the ability to add carbohydrate residues to the side-chains of amino acids) depending on the species, the cell type, and the culture conditions. Even the process of transfection can induce the expression of previously unexpressed glycosyl transferases (the enzyme which carries out glycosylation). Thus it is essential to assess the glycosylation pattern of glycoproteins, in particular when they are destined for therapeutic use.

10.3 What kinds of products are made using animal cells cultured *in vitro*?

recombinant
proteins

As we mentioned in Chapter 1, animal cells have been used commercially in the production of viruses for vaccines, and for the production of monoclonal antibodies. It was not long before pharmaceutical companies became interested in generating proteins of therapeutic value from genetically engineered cells. The manufacture of proteins from animal cells has become an important component of the biotechnology industry. Examples of recombinant proteins which are now produced on a large scale from mammalian cells in culture include tissue plasminogen activator (used in the treatment of thrombosis), the blood clotting factors such as factor VIII (used in the treatment of haemophilia) and factor IX (used in the treatment of Christmas disease), and erythropoietin (used in the treatment of anaemia - see also Chapter 8). In addition, genetic engineering of cultured cells has been important in generating new vaccines which are safer, have reduced side effects, and are more effective immunologically. The key to these improved vaccines was to produce only a viral protein rather than the whole virus, by expressing a viral gene product in cultured cells. In 1990 a hepatitis B vaccine was produced in this way using Chinese hamster ovary cells, and vaccines against influenza virus and HIV are in the experimental stages. Many of these products had previously been obtained from human blood provided by donors. However, it is often expensive to purify proteins from human blood, and there are risks of hepatitis and AIDS inherent in these products.

In this chapter we will not examine any single product in any great depth since we are primarily concerned with the principles of genetic engineering using animal cells. The application of these techniques in the production of sub-unit and attenuated strain vaccines and in the production of hormones (eg erythropoietin) are examined in a series of case studies in the BIOTOL text 'Biotechnological Innovations in Health Care'.

The genetic engineering of animal cells to produce recombinant proteins was dependent upon the development of recombinant DNA techniques - the ability to form different combinations of genetic material *in vitro*.

⊓ What other techniques would you need to genetically engineer these cells?

transfection of
animal cells

The advances in genetic engineering were also dependent upon the technology for introducing the manipulated genetic material into animal cells. This technology originated from the observations that naked uncoated protein-free viral DNA can enter cells and change the phenotype of cultured cells. Such transfer of genetic material was first achieved by Szybaska and Szybaski in 1962, although they were unable to demonstrate that cells had actually taken up DNA. Burnett and Harrington were the first to prove that cultured cells would take up DNA by treatment of cells with DEAE-dextran and viral DNA in 1968. The viral life cycle was initiated in a small proportion of these cells, and the cells were said to be transfected. The uptake of foreign DNA by animal cells has been called transformation, but to avoid confusion with the neoplastic transformation of animal cells referred to in Chapter 6, we will use the term transfection as a general term to describe this process. This chapter will cover the various different methods for introducing genes into animal cells which have been developed since these early demonstrations. You will need to be quite familiar with the basics of gene expression, and if you are not confident in this area, we recommend the BIOTOL books 'Infrastructure and Activity of Cells'.

10.4 Gene manipulation

We mentioned earlier that gene manipulation involves making changes in the DNA, and a number of basic skills are needed to do this. These methods are covered in great detail in other BIOTOL books, but briefly, fragments of DNA are inserted into a vector to produce a recombinant DNA molecule. The vector acts as a vehicle which transports the gene of interest into the host cell. The host cell divides and copies of the foreign DNA are produced making a clone of cells carrying the foreign gene.

The relevant techniques include:

- the preparation of pure samples of DNA;
- the cutting of DNA;
- the analysis of DNA fragment sizes;
- the joining together of DNA molecules;
- the introduction of DNA into bacteria for amplification;
- the identification of bacteria that contain the correct recombinant DNA molecule.

Most of these techniques are beyond the scope of this text, and we will concentrate only on some major points.

10.4.1 Isolation of the gene

In order to genetically engineer cultured cells, the gene of interest has first to be isolated, and this is not an easy task. For now, it is sufficient to know that a gene is frequently introduced as copy (complementary) DNA (cDNA) rather than as genomic DNA.

∏ In practice, what is the main difference between cDNA and genomic DNA?

You may remember that in eukaryotes, gene expression involves transcription, RNA processing, and translation, and that one of the events during RNA processing is the removal of introns from the heteronuclear RNA, which gives rise to mRNA. The mRNA can be used by the gene manipulator to create DNA by virtue of complementary base pairing, and this DNA is called cDNA (copy DNA). To achieve this we use an enzyme called reverse transcriptase. This synthesises single stranded DNA using RNA as a template. Strictly speaking, this enzyme is called RNA-dependent DNA polymerase. More usually, however, it is referred to as reverse transcriptase. A double stranded DNA molecule may be synthesised from this single stranded DNA by using suitable DNA-dependent DNA polymerase. The main difference between genomic DNA and cDNA is that cDNA has no introns, whereas genomic DNA contains introns. Some genes contain so many introns that they become unmanageable and are too large to incorporate successfully into a vector, whereas the corresponding cDNA is less problematic. By using cDNA, the cell which receives this DNA by transfection does not have to 'process' (ie remove introns) it in order to produce mRNA.

We have outlined this process in the form of a flow diagram associated with SAQ 10.2. There are some labels missing which you are asked to complete. This diagram provides you with a useful overview of the steps involved in introducing new genes (as cDNA) into new cell lines.

| **SAQ 10.2** | Fill in the empty boxes in the scheme below using the words provided below. |

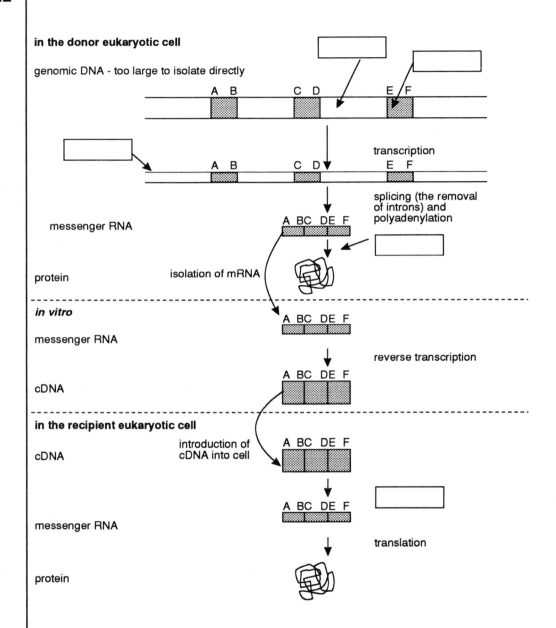

in the donor eukaryotic cell

genomic DNA - too large to isolate directly

A B   C D   E F

transcription

A B   C D   E F

splicing (the removal
of introns) and
polyadenylation

messenger RNA   A BC DE F

protein   isolation of mRNA

- -

in vitro

messenger RNA   A BC DE F

reverse transcription

cDNA   A BC DE F

- -

in the recipient eukaryotic cell

cDNA   introduction of   A BC DE F
              cDNA into cell

messenger RNA   A BC DE F

translation

protein

Word list
Heteronuclear RNA; translation; transcription; intron; exon.

10.4.2 Expression of the gene

In order to produce a desired protein in animal cells in culture, it is important to introduce the DNA in a form that can be expressed.

Π   What is meant by the term gene expression?

gene products

Gene expression refers to the synthesis of the final product of the gene. This product may be RNA, a protein, or a modified protein such as a glycoprotein. In the case of a protein product, the DNA has to be transcribed, and the transcripts have to be correctly processed, translocated to the cytosol, and subsequently translated. The polypeptide chain may need further processing, and the final product may either remain in the cell or be secreted. To optimise protein expression, all of these processes need to be considered. Apart from these requirements, the precise system used to express genetic information will vary with each task, and a number of choices regarding the type of expression system have to be made.

10.4.3 Stable or transient expression

A number of alternative techniques have been developed for introducing DNA into cells, and they fall into two groups:

- those designed for transient expression;

- those in which stably transfected cell lines are selected.

isolation of eukaryotic genes

In transient expression systems, DNA is introduced into cells and after 1-3 days the cells are harvested and the expression products of the introduced genes are analysed. Since the DNA has not necessarily integrated into the genome, the long-term culture of these cells results in dilution and eventual loss of the gene of interest from the cells. Transient expression is often used in research but is less relevant to the large scale production of specific proteins. One area of research, which is worth mentioning, is the isolation of eukaryotic genes using transient expression systems. The main aim of such experiments is to isolate a specific gene from the total DNA of a cell. This is quite a difficult task and there are a number of approaches depending on the type of information available about the gene of interest and the protein which it codes for.

expression cloning

If antibodies have been raised to the protein, or if the protein has suitable characteristics, then the transient cloning expression system is often used. Genomic fragments, or cDNAs, are incorporated into expression vectors which are subsequently introduced into cells in culture. The transiently transfected cells are screened using antibodies to detect the protein of interest, and only those cells which have taken up the expression vector carrying the right gene will show a reaction with the antibody. The gene can then be isolated and studied in more detail.

In stable systems a small proportion of the cells will retain the genetic information, either because it has integrated into the chromosomes, or because sequences facilitating DNA replication are included in the construct (a piece of DNA constructed by recombinant DNA technology). The newly introduced DNA is passed on to all subsequent generations. Since the uptake of DNA into animal cells lines is not always very efficient, it is important to select those cells which have incorporated the foreign DNA. We will come back to this point later.

10.4.4 Viral or bacterial plasmid vectors

vectors

expression
cassette

Ideally, DNA would be introduced into cells gently and efficiently without disrupting normal cellular processes. DNA can be carried into cells using vectors (literally 'carriers'), although as we shall see later, no ideal vector exists and in practice compromises have to be made. The cloning vectors used today are generally derived from either bacterial plasmids or viruses. As well as carrying the DNA into the cell, vectors have been constructed to allow gene expression by including a region called the expression cassette.

plasmid vectors

Plasmids are small pieces of circular DNA which occur naturally in bacteria and replicate independently of the bacterial genome. Genes of interest can be cloned into a bacterial plasmid, and the bacterial plasmid can be reintroduced into bacteria, where it replicates producing large amounts of the plasmid of interest. The simplest cloning vectors are based on the bacterial plasmid pBR322 (see Figure 10.1a), which has two important features.

selectable
marker

replication
origin

Firstly, it carries genes coding for antibiotic resistance (amp^R and tet^R), which allow for easy identification of bacteria which have taken up the vector. These are examples of selectable markers. Secondly, it carries a bacterial origin of replication (*ori*) which allows the plasmid to replicate independently of the bacterial genome. pBR322 allows manipulations to be carried out in bacteria and facilitates amplification of the recombinant plasmid in bacteria.

∏ Give two reasons why pBR322 may not be suitable as a mammalian cloning vector.

The main reasons why pBR322 is inadequate as a mammalian cloning vehicle are that the two features mentioned above, (the ability to select for plasmid uptake, and the ability of the plasmid to replicate), would not operate in mammalian cells. In addition, you may have mentioned a third valid point, that is, that DNA incorporated into pBR322 would not be expressed in mammalian cells because the machinery for gene expression in mammalian cells would not recognise the bacterial expression regions.

10.4.5 Mammalian vectors

Despite the points made in the previous section, the simplest mammalian cloning vectors are based on pBR322, but with a few modifications.

selectable
markers

Let us look first at the ability to select those cells which have taken up the plasmid. As we shall see later, some of the methods of introducing DNA into animal cells are not very efficient, and it is therefore essential to select those cells which have stably incorporated the foreign DNA. In the case of bacteria, the selection of those bacteria which had taken up vector DNA was based on the use of marker genes such as amp^R or tet^R, followed by growing the bacteria in the presence of the appropriate antibiotic (ampicillin or tetracyclin). A similar strategy is used in mammalian cells, except that the antibiotic G418 is used. Mammalian cells are naturally sensitive to an antibiotic called G418, although cells expressing the '*neo*' gene become resistant to this antibiotic. Cells which express the *neo* gene can inactivate G418.

The *neo* gene can be added to pBR322, as in the plasmid pSV2*neo* (shown in Figure 10.1b). Any cell which has taken up pSV2*neo* will be able to grow in medium containing G418. Similarly there is a gene called '*hph*' which enables mammalian cells to survive in hygromycin C, and may be used as an alternative to the *neo*/G418 system. Both G418 and hygromycin act by inhibiting eukaryotic translation. The drug selection protocol is followed by isolation and characterisation of individual transfectant clones. This is a labour intensive programme, requiring several weeks of tissue culture. Despite this drawback, the technique is commonly used, since the result is a stable expression system which can be used indefinitely.

What about the ability of plasmids to replicate in mammalian cells? We said earlier that plasmid based vectors have a bacterial origin of replication but we have not said anything about an origin of replication that facilitates DNA replication in mammalian cells. If a cell line is to produce a recombinant protein in a stable manner, then the gene of interest must be replicated in animal cells.

∏ State any possible reason why it may tricky to include a mammalian origin of replication in a cloning vector.

Again you will need an adequate knowledge of eukaryotic genetics to answer this question. In fact, the origins of replication in animal cells are poorly understood, and a discrete region which directs DNA replication has not been isolated. There are two tactics which are used to overcome this problem. Most plasmid-based systems depend on the integration of the genetic material into the mammalian genome, so that it is replicated along with the rest of the genome.

An alternative strategy has been to use the origins of replication of viruses which infect animal cells. These have been well characterised and facilitate replication of DNA in animal cells. These viral origins of replication are incorporated into plasmid vectors which are stably maintained as extrachromosomal elements at 10-100 copies per cell without integration. Such vectors are called shuttle vectors, since they can replicate in bacteria and in animal cells, for example the plasmid shown in Figure 10.1c contains the bovine papilloma virus origin of replication (for replication in mammalian cells) and

shuttle vectors   the pBR322 origin of replication (for replication in bacterial cells). There are also vectors which contain the SV40 origin of replication, and these cells can only replicate in the presence of a protein called the large T antigen. The gene for this protein can either be provided on the vector, or it can be provided by the cells, as in COS cells (which are derived from African Green Monkey kidney). WOP cells, on the other hand, express

T antigens   polyoma virus T antigen and can be used in conjunction with expression vectors containing the polyoma origin of replication. Expression vectors based on adenovirus and vaccinia virus also exist.

Sometimes researchers want to have many copies of newly introduced DNA in order to obtain higher levels of gene expression, and a large number of amplification systems is now available. The system most commonly used is based on the *dhfr* gene coding for

amplification systems   the enzyme dihydrofolate reductase (DHFR) which catalyses the conversion of dihydrofolate to tetrahydrofolate. The cytotoxic drug called methotrexate binds to the DHFR enzyme, but mammalian cells can develop resistance to it by amplifying the *dhfr* gene. More enzyme is produced so there will still be enough active enzyme after all the methotrexate is bound. After long-term culture of the cells in methotrexate, there may be up to 2000 copies of the *dhfr* gene per cell. Moreover gene amplification also occurs in cells which have taken up a vector containing *dhfr*, and any DNA integrated next to the *dhfr* gene will also be amplified. Expression vectors using the *dhfr* gene have been developed (see Figure 10.1d) and can be used in many cells types, although they are particularly effective when used in conjunction with *dhfr*-deficient Chinese hamster ovary cell lines.

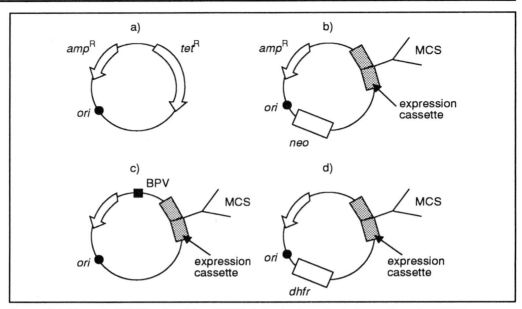

Figure 10.1 Maps of pBR322 and various pBR322-derived eukaryotic cloning vectors. a) pBR322 - a commonly used bacterially derived plasmid. b) pSV2*neo* is derived from pBR322 by addition of the *neo* gene which allows selection of transfectants using the antibiotic G418, and an expression cassette derived from the SV40 virus. c) A shuttle vector containing the bovine papilloma virus origin of replication (for replication in mammalian cells) and the pBR322 origin of replication (for replication in bacterial cells), as well as an expression cassette. d) pSV2*dhfr* is a pBR322-based amplification vector, again it contains the expression cassette derived from the SV40 virus, and in addition the *dhfr* gene which facilitates DNA amplification (see text for further discussion). amp^R = ampicillin resistance gene, tet^R = tetracyclin resistance gene, *neo* = neomycin phosphotransferase gene, *ori* = bacterial origin or replication, MCS = multiple cloning site; the gene of interest is inserted into this site. BPV = Bovine Papilloma Virus origin of replication, shaded boxes = expression cassettes.

| SAQ 10.3 | Examine Figure 10.1 carefully, noting the various genes carried by each plasmid. Why do you think all of the illustrated eukaryotic expression plasmids still carry the ampicillin resistance gene? |

SAQ 10.4

Below is a selection of potential mammalian vectors. Examine them carefully.

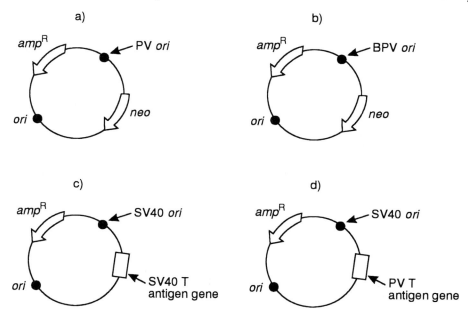

ori = bacterial origin of replication, PV ori = polyoma virus origin of replication, BPV ori = bovine papilloma virus origin of replication, SV40 ori = SV40 origin of replication, neo = resistance to G418.

Which of these will replicate in animal cells? (consider each separately).

10.4.6 Expression vectors

Most useful plasmid-based mammalian vectors contain an expression cassette. Whatever vector is used, the coding region of the gene must be adjacent to control regions which animal cells recognise. These regulatory sequences are different from those found in bacteria. The DNA elements necessary for eukaryotic gene expression have been studied extensively and information about these DNA elements has been exploited to produce a battery of gene transfer vectors. We will briefly consider the various DNA elements that control eukaryotic gene expression, and the use of these elements in the expression cassette.

Promoters

At the 5′ end of eukaryotic genes are the promoters and enhancers which control the level and tissue specificity of DNA transcription. Early studies demonstrated that the structure of promoter elements is relatively conserved, although there are differences between the promoters used by different types of RNA polymerase. We will limit the discussion to the genes coding for proteins which are transcribed by RNA polymerase II (pol II).

One critical aspect of all pol II promoter elements is that they contain mRNA cap sites - the point at which the mRNA transcript actually begins. By convention the transcription start site is +1, and everything 5′ or upstream of this is assigned a negative number, whereas everything 3′ or downstream is assigned a positive number. Upstream of the cap site is a frequently encountered sequence known as the TATA box.

This centres around -25 (that is 25 base pairs upstream of the mRNA cap site) and is important for locating the start site of transcription. Further upstream, the relative positions of the promoter elements become more variable. The CCAAT box resides roughly 70-80 base pairs upstream from the mRNA cap site and is important for the initial binding of RNA polymerase II. Another promoter element of many housekeeping genes consists of multiple copies of GC rich regions upstream of the mRNA cap site. There appears to be an absolute requirement for promoter elements in eukaryotic gene transcription, so they are included in all recombinant plasmids.

 enhancers

In addition to promoter elements, other DNA consensus sequences are involved in the expression of genes in eukaryotes. They are called enhancers because they enhance the transcription of genes, and they were first identified in the genomes of viruses. An enhancer is a gene sequence that is required for the full activity of its associated promoter. It needs not have a fixed position or a fixed orientation relative to the gene. Enhancers are important for tissue specific gene transcription.

Note that the molecular organisation of eukaryotic genes is described in detail in the BIOTOL text 'Genome Management in Eukaryotes'.

SAQ 10.5

Using the list provided, assign the correct labels to the shaded areas in the diagram below.

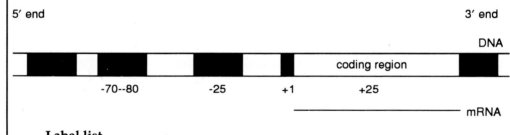

Label list
mRNA cap site; TATA box; CCAAT box; GC box; enhancer.

The exact enhancer used depends on the specific cell type in which expression is to take place and on whether expression is to be constitutive or inducible.

∏ Give a reason why it might be particularly useful in the biotechnology industry to be able to switch genes on and off.

If the protein products have a harmful effect on the cell then there would be a selection against stable cell lines constitutively expressing such a product. However, using an inducible system, large amounts of cells can be grown up and then induced to produce the protein of interest, which can then be purified. A common approach has been the use of heat shock promoters such as the *Drosophila hsp* 70 promoter, which only directs expression of the adjacent coding region at 43°C.

heat shock promoters

Many different expression vectors have been developed using the coding portions of one gene and the regulatory portions of another gene. For example, a gene which is expressed at low levels can be combined with the promoter of a gene which is expressed at high levels in order to increase the level of expression of a protein. Many promoter regions have been taken from viruses. The retroviral long terminal repeats (LTRs), the

SV40 early regions, and the adenovirus major late promoter are only a few examples of promoters which are frequently used in gene manipulation.

At the 3′ end of the gene, sequences necessary for termination of transcription, cleavage of the 3′ end of RNA, and polyadenylation have to be present. Sequences which confer stability on the RNA transcript may be added to the recombinant plasmid. The recognition sequence for translation must also be present, and sequences which enhance the interaction of RNA with the ribosomal initiation complex may increase the rate of translation.

| **SAQ 10.6** | Identify the genetic elements you might use to express a gene in animal cells from the list below. (If your knowledge of molecular biology is limited, you might find it difficult to decide about some of the examples included. In this case, read our response carefully). |

1) The SV40 early region   2) Polyadenylation signals

3) Promoter   4) Proteins

5) The Shine-Dalgarno sequence   6) The recognition sequence for translation

7) Secretion signals   8) The Pribnow box

9) Enhancer   10) cDNA

11) *dhfr*   12) *amp*^{R}

10.4.7 Expression cassettes

We will now look at two examples of expression cassettes in more detail. Firstly, the pSV2 plasmids shown in Figure 10.1b) and d) contain a fragment of SV40 DNA which includes the promoter for early transcription and the transcriptional start point. Then comes a sequence containing a small intron, followed by the polyadenylation site. The SV40 expression signals function in a wide variety of mammalian cells, providing a generally useful expression cassette, shown in more detail in Figure 10.2a. The long terminal repeat (LTR) of mouse mammary tumour virus (MMTV) has been particularly useful as an inducible system, since transcription is activated only in the presence of steroid hormones. The 3′ signals from the SV40 virus can be used in this expression cassette too, as shown in Figure 10.2b.

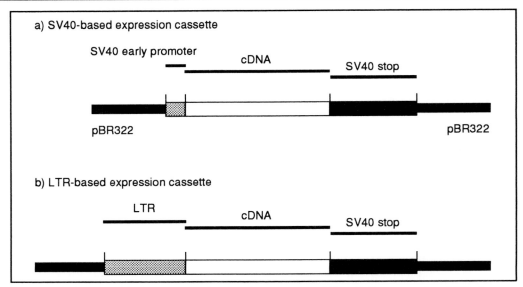

Figure 10.2 Diagram of 2 different types of expression cassettes. a) Constitutive expression. SV40 stop refers to the splice and polyadenylation sites from SV40. cDNA indicates the site where the cDNA of interest is incorporated. b) Inducible expression. LTR denotes the long terminal repeat of MMTV.

SAQ 10.7

Match the genetic elements in the list on the left to their functions in the list on the right.

1) *amp*^R	allows for amplification of genes
2) *neo*^R	provides a means of selecting eukaryotic cells that have received foreign DNA
3) SV40 early region	provides a means of selecting bacterial cells that have received foreign DNA
4) LTR	facilitates inducible expression of genes in eukaryotes
5) *dhfr*	facilitates constitutive expression of genes in eukaryotes

10.4.8 Retroviral vectors

Alternatives to plasmid-based vectors are retroviral vectors. Retroviruses are RNA viruses that replicate by copying RNA to DNA, inserting the DNA into the host cell genome, and then transcribing the DNA. Thus they have an RNA genome, but integrate a DNA copy of this RNA into the chromosomes of the cells they infect. Most retroviruses do not kill their host cells, but viral particles continually bud from the surface membrane of infected cells.

retroviruses integrate into host cell's genome

Retroviruses are very powerful tools in tissue culture since they have effective systems of infecting cells, they often contain powerful promoters, and their DNA integrates into the host cell genome as part of the normal retroviral replication cycle. These properties can be exploited by incorporating genes of interest into retroviruses and then using the genetically engineered retroviruses to infect cells. This is the most efficient method of introducing genes into cells and therefore deserves a more detailed description. We will start by giving some more information about the life cycle of the retrovirus.

Use Figure 10.3 to help you to follow the description given below.

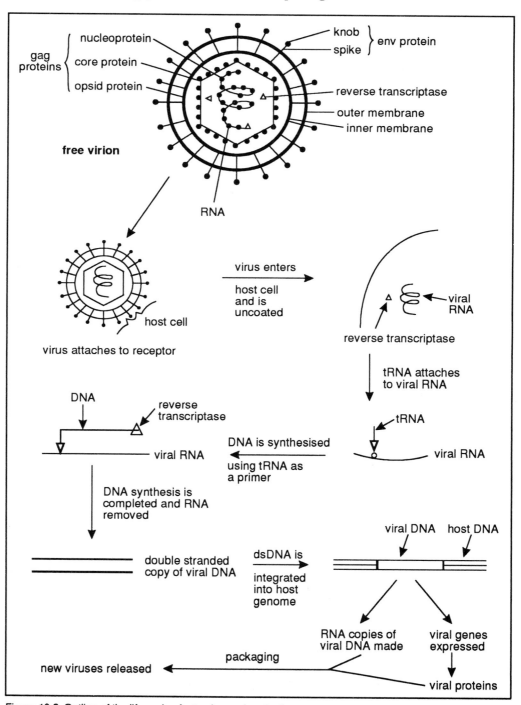

Figure 10.3 Outline of the life cycle of retroviruses (see text).

Normal life cycle of a retroviruses

Retrovirally infected cells shed membrane encapsulated viruses. Each viral particle containing two molecules of RNA. The virus, studded with an envelope glycoprotein (which serves to determine the host range of infectivity), attaches to a cellular receptor in the plasma membrane of the cell being infected. After receptor binding, the virus is internalised and uncoated as it is passed through the cytoplasm of the host cell. Reverse transcriptase molecules, resident in the viral core, catalyse the synthesis of double stranded DNA provirus. This process is primed by the binding of a tRNA molecule to the genomic viral RNA and the double stranded DNA provirus is subsequently integrated into the genome of the host cell where it serves as a transcriptional template for both mRNA encoding viral proteins and virion genomic RNA. These are packaged into viral core particles. On the way out of the infected cell, core particles move through the cytoplasm and attach to the inside of the plasma membrane of the newly infected cell. As they bud they take with them tracts of membrane containing the virally encoded 'env' glycoprotein gene product.

The description of the events of the life cycle of retroviruses given above is somewhat simplified and does not include a description of the events surrounding the synthesis and integration of double stranded DNA using the viral RNA as a template. This is a fascinating process and well worth discussion.

The RNA molecules of the virus contain terminal nucleotide sequences that are repeats of each other. They also contain an inverted repeat sequence. We can represent this in the following way:

$$3' \text{ poly A} \underline{\hspace{1cm}}_{\text{ABC}} \underline{\hspace{4cm}}_{\text{C'B'A'ABC}} 5'$$

The tRNA attaches to this RNA molecule.

$$3' \text{ poly A} \underline{\hspace{1cm}}_{\text{ABC}} \underline{\hspace{4cm}}_{\text{C'B'A'ABC}} 5' \quad \text{tRNA}$$

and DNA synthesis begins using the enzyme reverse transcriptase.

$$3' \text{ poly A} \underline{\hspace{1cm}}_{\text{ABC}} \underline{\hspace{4cm}}_{\text{C'B'A'ABC}} 5' \quad \text{DNA}$$

An RNase (ribonuclease H) then hydrolyses a few nucleotides from the 5' end of the RNA, to produce the structure shown below:

$$3' \text{ poly A} \underline{\hspace{1cm}}_{\text{ABC}} \underline{\hspace{4cm}}_{\text{C'B'A'}}$$

The second copy of the viral RNA then hybridises (hydrogen bonds) with the DNA and DNA synthesis continues.

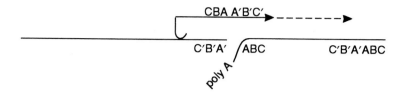

Ribonuclease H then digests away the RNA to leave ssDNA.

$5'$ $\underline{\text{CBA A'B'C'} \qquad\qquad \text{CBA A'B'C'}}$ $3'$

The 3' end of the molecule folds back on its self.

$5'$ CBA A'B'C' C B A
C'B'A'

and DNA synthesis continues.

C B A A' B' C'   C B A
C'B'A' A B C   C'B'A'

An endonuclease cuts the loop to form a double stranded DNA molecule.

$5'$ $\overline{\text{C B A A' B' C'} \qquad\qquad\qquad \text{C B A}}$ $3'$
$3'$ $\underline{\text{C'B'A' A B C} \qquad\qquad\qquad \text{C'B'A'}}$ $5'$

The next question is, how does this DNA become integrated into the host genome?

It is thought that a 3' exonuclease digests a few nucleotides from either end of the dsDNA molecule. Thus:

CBA A'B'C'

A B C   C'B'A'

The two ends of the molecule are now complementary so the molecule can then cyclise.

This molecule then integrates via a complementary sequence into the host's genome.

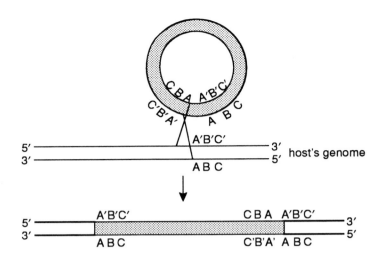

The upper strand is transcribed to produce viral RNA.

$$3' \underline{\text{ABC} \qquad\qquad \text{C'B'A' ABC}} \; 5'$$

This elegant mechanism of viral replication and integration is well worth remembering. It is an excellent illustration of how repeat (and inverted repeat) sequences of nucleotides enable nature to manipulate nucleotides. It also provides a mechanism by which new genes (in this case retroviral genes) may be inserted and integrated into host genomes. In principle, we can hijack this system to introduce new genes into host cells.

In order to do this, we will need to know a little more about the genes in the retroviral genome. We will describe these in a little more detail in the next sub-section.

The retroviral genome

In order to construct retroviral vectors we need to know about the sequences that are essential for integration and replication of the retroviral genome.

∏   From our earlier description, what genes or nucleotide sequences are essential for the production of viral DNA and its integration into the host genome?

We anticipate that you would have decided that the terminal repeated sequences and inverted repeat sequence are essential. For simplicity, we will refer to them simply as long terminal repeat (LTR) sequences. You may also have included a binding site for the primer (tRNA). This is called the Primer Binding Site (PBS). You should have also predicted that the genome of the virus would need to code for the enzyme reverse transcriptase. This gene is usually called *pol* from polymerase.

Thus from what we have already discussed you could construct a list of essential genes or nucleotide sequences. These are LTRs, PBS and *pol*.

There are, however, other genes. These include *gag* which codes for the viral core proteins and *env* which codes for the envelope proteins.

There is also a short sequence which is necessary for the efficient packaging of the viral RNA into virions, this sequence is called the packaging signal (ψ).

We need also to mention two sites called 5'ss and 3'ss which are used for the production of sub-genomic mRNA (ie RNA that does not extend over the whole length of the genome). These are the so-called viral splice donor and acceptor sequences.

The actual arrangement of these genes and sequences in a typical retrovirus is shown in Figure 10.4.

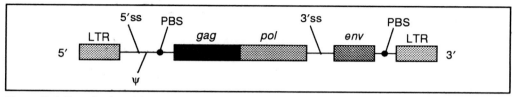

Figure 10.4 Organisation of the genome of a typical retrovirus.

Although all of these regions are essential for retroviruses, not all are needed to create useful vectors for genetic engineering.

<table>
<tr><td>

SAQ 10.8

</td><td>

Anticipate the outcome of the following.

1) A retrovirus infects a suitable host cell, but its ψ sequence has become deleted.

2) A retrovirus has been constructed such that its *gag* gene has been deleted. This virus is used to infect host cells.

3) A retrovirus with no *pol* gene is allowed to co-infect host cells with intact retroviruses.

4) A retrovirus with no *env* gene co-infects a cell with intact retroviruses.

</td></tr>
</table>

Engineering retroviruses

Retroviruses can be engineered so that they only contain the sequences necessary for integration into the genome. The strategy used is to strip the virus of the dispensable *gag, pol* and *env* genes, thus generating a replication defective virus which requires other viral functions in order to replicate. The only things which are retained are:

- the two LTRs (which facilitate integration);

- the primer binding sites;

- the packaging signal (ψ).

Many laboratories have shown that retroviruses containing foreign genes can be obtained in high titres. In 1982, a gene (the Herpes Simplex Virus thymidine kinase gene) was successfully integrated into the Moloney murine leukaemia provirus. Proviral DNA containing the thymidine kinase gene was transfected into host cells, but the virus could not reproduce itself in the normal manner since some of its essential functions were now missing (the virus was defective). These functions could be replaced using a helper virus, and the recombinant retroviral DNA was packaged into viral particles and secreted into the medium. The gene was expressed upon subsequent infection of cells with this recombinant retrovirus.

Retroviral vectors can be constructed using only the LTRs and immediate flanking sequences of the retroviral genome, and retroviral vectors with other useful features have been developed. In 1984 a retroviral vector called pZIPNEOSV(X)1 was developed from a murine leukaemia virus and used to drive the expression of two inserted genes - the *neo* gene (included as a selectable marker) and the gene of choice. The expression of both genes can be driven from the 5' LTR. The vectors are used in combination with φ2 and φAM packaging lines. (Packaging lines are viruses that are used to co-infect cells and which will enable replication and packaging of recombinant retroviral vectors. For example if the vector carries no *pol* or *env* genes, it will only be replicated and packaged in cells containing viruses carrying these genes). Cells integrating the retrovirus can be selected using G418 (see Figure 10.5). MCS is a multiple cloning site, as it contains nucleotide sequences which allow us to use a variety of restriction enzymes to cut the vector in this region. New genes can then be inserted at this cut site.

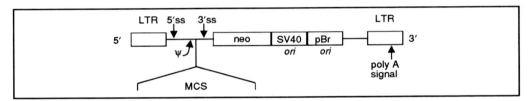

Figure 10.5 Map of the recombinant retroviral vector pZIPNEOSV(X)1 (see text for discussion).

10.4.9 Integration sites

Mammalian cells generally integrate newly introduced DNA in a non-specific manner. However, for several reasons it is desirable to introduce the gene at precise chromosomal locations.

∏ Suggest as many reasons as you can for the above statement.

gene targeting

The insertion of a DNA sequence into a specific site in the genome is often called gene targeting. More specifically the method used is called homologous recombination, since it invariably involves the combination of certain DNA sequences in the chromosome with homologous sequences present in the newly introduced DNA molecule. The opposite of gene targeting is random integration and one obvious disadvantage of random integration is that it may disrupt another important gene.

Another reason for gene targeting is that there seem to be certain locations in the genome which are expressed at higher levels than others and some regions are transcriptionally silent, although the reasons for this are poorly understood. For this reason the site of integration will affect the efficiency of recombinant protein production. Thus if the gene is integrated into a transcriptionally active site, the encoded protein may be expressed at higher levels than those obtained by random integration. Preliminary results indicate that expression can be increased several-fold by taking this into account.

A third reason for gene targeting is that it enables endogenous genes to be modified whereas previously it was only possible to incorporate an exogenous gene into the genome leaving the endogenous gene in tact. Although gene targeting has been used for several years in yeast, it has only recently become possible in mammalian cells.

ES-cells

One of the largest areas of activity of gene targeting is in its application to the embryonic stem cells (ES-cells) which were mentioned briefly in Chapters 1 and 9. The embryo is removed during the blastocyst stage of development when the cells are still pluripotent. Cells can be isolated and grown in culture for many generations and are of particular interest because they can be manipulated in culture and then re-introduced into mouse embryos (see Figure 10.6). The animals which grow from these embryos are said to be chimaeric (chimeric), since they consist of cells derived from the genetically manipulated ES-cells as well as normal cells. (The chimaera is a mythological monster consisting of the head of a lion, the body of a goat, and the tail of a serpent. In biology it has come to mean an organism that consists of at least two genetically different kinds of tissue). The next generation of animals will either consist entirely of normal cells or entirely of cells with the desired change (and these animals are called transgenic since they contain the artificially introduced gene). Thus ES-cells can be used to construct a whole organism with the specific mutant allele. Often mice are used in these experiments.

chimaeric
animals

Π Why will the second generation of animals derived from chimaeric animals be composed of either 'normal' or 'changed' cells but not of a mixture?

You should have realised that if the 'changed' cells introduced into the blastocyst became germline cells (ie sperm or ova producing cells), then the gametes from such an animal may contain such 'changed' genomes. Thus zygotes obtained from these gametes may also contain this 'changed' genome. If on the other hand, the germ line cells in the chimaera were derived from 'normal' cells, none of the offspring would receive the genetically modified genomes.

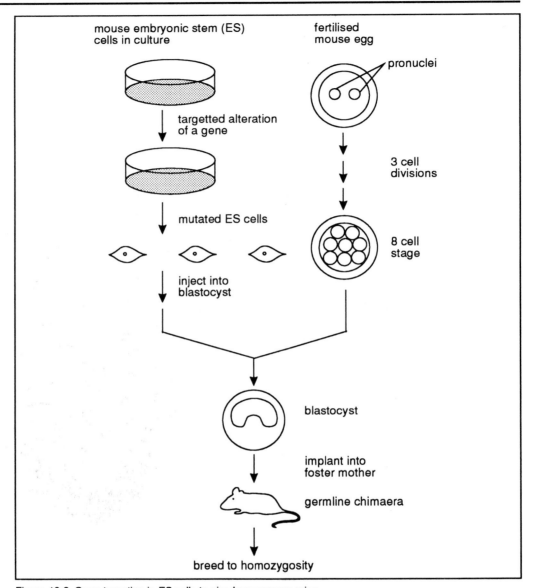

Figure 10.6 Gene targeting in ES cells to give homozygous mice.

∏ If genetically changed cells are introduced into a blastocyst and the introduced cells become germ line cells, are the offspring of such an animal guaranteed to carry the genetic change?

The answer depends upon whether or not the genetically changed cells are homozygous or heterozygous with respect to the genetic modification.

We can illustrate this in the following way:

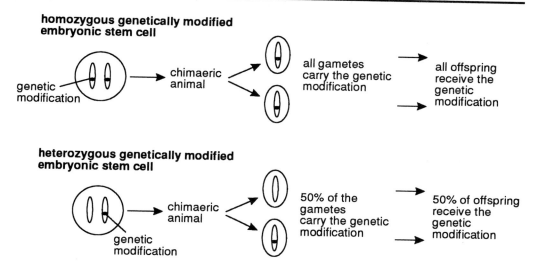

homozygous genetically modified embryonic stem cell

genetic modification → chimaeric animal → all gametes carry the genetic modification → all offspring receive the genetic modification

heterozygous genetically modified embryonic stem cell

genetic modification → chimaeric animal → 50% of the gametes carry the genetic modification → 50% of offspring receive the genetic modification

The frequency of homologous recombination is still several orders of magnitude lower than the frequency of random integration so it is always necessary to select and screen transfected cells lines for homologously integrated DNA. The main advance in this area has been the development of selective enrichment strategies and rapid screening schemes using PCR (the polymerase chain reaction).

10.5 Introduction of the recombinant DNA

In this section, we will briefly review the ways in which recombinant DNA may be introduced into animal cells. For a fuller description of these techniques, we recommend the BIOTOL text 'Strategies for Engineering Organisms'. Here we will divide our discussion into:

- retroviral infection;

- calcium phosphate transfection;

- protoplast fusion;

- lipofection;

- electroporation;

- microinjection;

- microprojectiles (biolistics).

In the final part of this section, we will briefly consider the formation of concatemers from DNA introduced into cells and its consequences on selection strategies.

10.5.1 Retroviral infection

Retroviruses were obvious candidates for use as vectors since the DNA of retroviruses becomes incorporated into the host cell genome. Many useful vectors have been derived from retroviruses, in particular avian retroviruses such as Rous sarcoma virus (RSV) and murine retroviruses such as Moloney murine leukaemia virus (M-MuLV). These vectors were discussed in detail in Section 10.4.4.

In the retroviral vectors, the gene of interest replaces some of the viral genes which are needed for replication so the recombinant vector has to be introduced into a very special cell line which provides the missing retroviral functions (called packaging cell lines). The engineered retroviruses are produced in the packaging cell lines and the secreted (defective) virus from these cells is used to infect the target cells, where integration and expression occur efficiently, without the risk of producing infectious virus.

Despite their efficiency at introducing DNA, viral vectors have a number of disadvantages. Viral infection sometimes changes the cell's properties, for example epithelial cells may lose their cell surface polarity. In addition, certain retroviruses are acutely oncogenic which precludes their use in a number of types of work. Retroviruses are also prone to delete gene sequences, and they can also exchange gene sequences with other retroviruses. Thus a retroviral vector could combine with an endogenous viral sequence to produce infectious recombinant viruses.

Recently a new vector system based on viruses has been developed. DNA is attached onto the outside of the virus which penetrates the cell via its normal route using surface receptors. It enters the nucleus via normal cellular uptake and transport processes. The advantage of this vector is that it can transport up to 48 kb of DNA successfully. However, this vector is still in the experimental stage of development.

10.5.2 Calcium phosphate transfection

The method which is most commonly used to introduce DNA into cells, is based on co-precipitation of DNA with calcium phosphate first described for the infection of cells with adenovirus 5 by van der Eb and for the transfer of cellular genes by Wigler. This method is more commonly known as transfection (although we are using the word in a more general sense in this text). The DNA is dissolved in a buffer containing calcium chloride, and an equal volume of buffer containing phosphate is added. A DNA precipitate is allowed to form at room temperature and this is added directly to the cells. The cells are incubated with DNA-calcium phosphate co-precipitate overnight and the cells are then placed in selective media to isolate stable cell lines.

It is not entirely clear what the function of the calcium is, although it probably concentrates DNA on the cell membrane to further facilitate uptake of DNA. In addition calcium protects DNA from hydrolytic nucleases. The co-precipitate sediments onto cells and becomes adsorbed onto the cell membrane, where it is taken up by (phagocytosed by) the cells. A small proportion of the DNA becomes stably integrated into the nuclear genome.

Calcium phosphate transfection is still frequently used for mammalian systems where possible. However, some cells (for example bone marrow cells) do not like having the solid calcium phosphate precipitate adhering to them, and some cells are simply hard to transfect using this method. There are several other techniques for introducing DNA into these cells.

10.5.3 Protoplast fusion

Another approach toward achieving high efficiency gene transfer has been to use direct fusion of cells to bacterial protoplasts. The bacterial cell wall is removed using lysozyme, and the bacterial protoplast is fused to the eukaryotic cell using polyethylene glycol. The main advantages of this method are that you do not have to purify the DNA from bacteria before introducing it into animal cells, and that the uptake of DNA is much more efficient, with a success rate of up to 100%.

10.5.4 Lipofection

Another method of gene transfer which relies on the fusion of membrane systems is lipofection. DNA in solution complexes efficiently with cationic lipids which form a single bilayer vesicle around the DNA called a liposome. When these liposome vesicles are added to eukaryotic cells, they seem to fuse with the plasma membrane and the DNA is taken up by the cell.

10.5.5 Electroporation

The cell membrane can be made permeable by exposing cells to a brief electric shock of several thousand volts. High voltage electric pulses make transient holes in the plasma membrane and any DNA which is present in the cell suspension can be taken up through these holes. The efficiency of this method may be very high. The exact conditions for electroporation, however, vary for different cell types so a lot a preliminary experiments have to be done to optimise the conditions. Thus this method is usually used only for cells which cannot take up DNA by transfection, such as myeloma cells.

10.5.6 Microinjection

Yet another method of introducing DNA into cells is by microinjection. Small volumes of DNA (10-100 fl, that is 10^{-8} to 10^{-7} µl) can be microinjected into the nucleus of tissue culture cells using a glass micropipette with the aid of a micromanipulator, as shown diagrammatically in Figure 10.7. Amphibian oocytes are popular targets for this method because of their large nuclei, but this method is not used very often for mammalian cells since it requires expensive and sophisticated equipment and considerable expertise. Obviously, cells have to be microinjected one at a time, and this is a major disadvantage of this method.

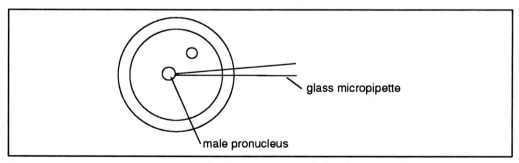

Figure 10.7 Diagram to show the microinjection of DNA into fertilised mouse oocytes using a glass micropipette. DNA is injected into the pronucleus of a newly fertilised mouse egg.

10.5.7 Microprojectiles (biolistics)

The most novel mechanism of introducing DNA into cells in culture is called biolistics (which stands for biological ballistics), and this method is used especially when the more standard techniques do not work for a particular cell line. The method uses pressurised helium to shoot tungsten or gold particles coated with DNA into the cells. Although this method is still in its infancy, it is likely to become an important mechanism of transfection.

10.5.8 Concatemer formation and selection

When cells take up foreign DNA, whether by calcium phosphate precipitation, electroporation, or microinjection, the DNA seems to form concatemers before it is integrated into the genome.

We remind you that concatemers are linearly repeated units of DNA. We can represent this process thus:

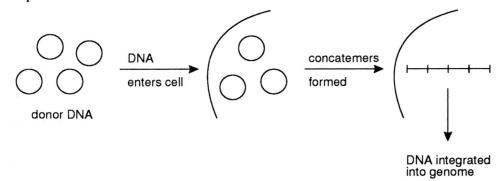

∏ What would happen if we used a mixture of plasmids?

We anticipate that you would predict that heterogenous concatemers would be produced. Thus:

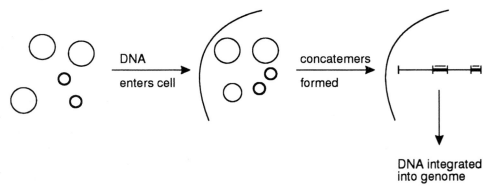

Thus selectable marker genes do not have to be included in the expression vector but can be added separately during the introduction of the genetic material. The transfection of the DNA of interest plus genes conferring resistance to antibiotics is known as co-transfection. An excess of the gene of interest is used so that most of the cells expressing the selectable marker will also be expressing the gene of interest. A typical co-transfection experiment would use 20 μg plasmid carrying the gene of interest and 1 μg of the resistance plasmid. Two days after transfection cells are split 1:5 in selective medium containing either 1 mg/ml G418 or 200 ng/ml hygromycin B. Visible colonies are isolated after 2 weeks, expanded, and tested for the presence and expression of the desired construct.

SAQ 10.9	Draw a flow chart outlining the steps for introducing DNA into cells by calcium phosphate transfection, starting with the vector and the gene of interest, and finishing with a stably transfected cell line.

10.6 The host system

We started this chapter by saying that although bacterial expression systems are extremely effective and economical, they cannot always produce the correctly folded biologically active proteins when the molecule of interest is large and complicated in structure. So far we have not mentioned which cells are chosen for the expression of such proteins. The choice of cell type into which recombinant genes are transferred has been based on technical considerations. Continuous cell lines vary considerably in their ability to take up and express foreign DNA, and a number of systems are useful for the production of eukaryotic proteins with the correct modification. We will now examine some of the commonly used cell lines.

Myeloma cells

Myeloma cells are the malignant counterparts of immunoglobulin secreting plasma cells. These cells are specialised at secreting and they have the machinery to add and modify sugar residues such as those found in mammalian glycoproteins. For many years, myeloma cells have been used extensively to generate antigen specific hybridomas. The genetic elements required for high immunoglobulin expression in myeloma cells have been well defined, and vectors have been designed containing the proper promoters and enhancers. These cells have been the natural choice for the production of modified immunoglobulins.

Insect cell lines

Insect cell lines provide a potentially important alternative to mammalian cell lines for the production of animal proteins with the correct post-translational modifications. Insect cells can perform a range of post-translational modifications which confer the correct folding, biological activity, and antigenicity on expressed proteins. Glycosylation occurs at the correct sites, although the more complex forms of N-linked glycosylation involving trimming of terminal sugars and addition of other sugars are not found. Thanks to a natural expression system they can provide extremely high yields of recombinant protein. One particularly useful expression system is based on the baculoviruses which infect insect cells but not vertebrate cells.

Baculovirus

The baculoviruses have a large double stranded circular DNA genome, which includes a gene coding for a protein called polyhedrin. This protein represents upto 50% of the total protein in the cell. During normal infection, the viruses produce nuclear inclusion bodies which consist of virus particles embedded in a matrix of polyhedrin. Transcription of the polyhedrin gene is driven by an extremely active promoter which is ideally suited for driving the expression of foreign genes. The gene of interest can be inserted just downstream of the polyhedron promoter, since the polyhedrin gene products are dispensable.

A large number of foreign genes from bacteria, plants, and animals have been expressed using baculoviruses and a number of baculovirus vectors have been constructed. High levels of expression can be obtained upon infection of cultured insect cells with these recombinant baculovirus vectors. A popular cell culture system uses cell lines (such as the Sf9 cell line) derived from the fall armyworm, *Spodoptera frugiperda*. Sf9 cells can be cultured at 20-28°C in insect cell culture media which are commercially available. Upon infection of Sf9 cells with the recombinant plasmids high levels of expression of the foreign gene are achieved.

| SAQ 10.10 | Now we will test your memory! A large number of cell lines have been described in this text. Many of these are particularly appropriate for genetic manipulation. See how many you can recall and comment on why they are particularly appropriate. We have given you examples to explain what we mean. |

Cell line	Why appropriate
Xenopus oocytes	Useful for micro-injection of DNA or RNA since cells are very large. Transient expression of DNA or RNA can be used to isolate genes.
Sf9 cells	Derived from the fall armyworm and used for baculovirus infections.

This is a rather open ended question. So try it for a while and then read our response.

10.7 The characterisation of transfected cells

We have covered a number of methods of introducing DNA into animal cells, although some of the more commonly used methods are not very efficient. It is therefore essential to identify those cells which have stably incorporated the foreign DNA and are able to express it.

∏ How would you determine whether the vector of interest has been incorporated into the cell?

Southern blotting   One technique to examine whether specific DNA sequences are present is Southern blotting.

The most common way this is done is to extract the DNA from the test material, cut it with a restriction enzyme and separate the fragments using gel electrophoresis. The separated fragments are then 'blotted' onto a membrane and melted to form single stranded DNA. These are then challenged with labelled DNA probes. If sequences complementary to those of the probe are present in the blotted DNA fragments, then the probe will form hydrogen bonds with the sequence-bearing fragments.

In some cases, we are not merely interested in whether or not foreign DNA has been incorporated into a host cell line, we would like to know if it is being expressed.

∏ What can you assay to prove that a gene is being expressed?

You really have two main options, either analyse the RNA transcript or the protein produced from such a transcript.

Northern blotting
The presence of particular RNA sequences in cells can be assayed in a manner similar to that used for DNA except that in this case it is called Northern blotting.

RNA is extracted from the cell and separated according to size by gel electrophoresis. Then this RNA is blotted onto a membrane which is then challenged with labelled probes which will hybridise with the appropriate RNA molecules. You should note that this is not the only technique available for detecting the presence of specific RNA molecules. The development of polymerisation chain reaction (PCR) techniques provides a method for making multiple copies of nucleotide sequences *in vitro*. This technique is, therefore, valuable in enabling us to produce copies of sequences either to provide sufficient test material in hybridisation studies or to produce the radioactive probes.

Western blotting
The detection of specific protein either depends upon measuring a particular activity of the protein (eg enzymatic activity) or its ability to bind particular antibodies. In this latter case, the cellular proteins are usually separated by electrophoresis and blotted onto a membrane. Then these membranes are challenged with labelled antibodies. Often the labels used are fluorescent dyes (hence immunofluorescence) or they may be tagged with radioactive isotopes. This technique of electrophoretically separating proteins, blotting them and allowing them to react with antibodies is called immunoblotting or Western blotting.

These techniques and related procedures, are described in depth in the BIOTOL text 'Analysis of Amino Acids, Proteins and Nucleic Acids' and also in 'Techniques for Engineering Genes'.

SAQ 10.11	A ^{32}P-labelled DNA probe containing the nucleotide sequence of a gene has been prepared. DNA collected from a cell thought to contain the gene, is subjected to the Southern blotting procedure. One of the fragments of DNA is shown to be capable of hybridising with the ^{32}P-labelled DNA probe. RNA, collected from the same source, is subjected to the Northern blotting procedure. The same ^{32}P-labelled DNA probe is used. In this case, the probe did not hybridise with any RNA fraction.
	Give two possible reasons why this result might have been observed. (Assume that control experiments show that the DNA probe is capable of hybridising both with the DNA from the gene and with its transcription product).

In addition to the analysis of the cells with respect to the integration of recombinant DNA, it may be necessary to confirm that the correct cell type has been transfected or infected. This is particularly important for transfection techniques which may involved many weeks of selection. The methods outlined in Chapter 4 may be applied to transfected cell lines.

SAQ 10.12	Identify from the list below the factors which influence the efficiency of production of a recombinant protein.
	1) Integration site 2) Origin of gene
	3) Copy number 4) Orientation of enhancer
	5) Transcriptional efficiency 6) Number of chromosomes in host genome
	7) RNA stability 8)RNA processing

SAQ 10.13	Put in the right order the steps which need to be carried out in order to produce a recombinant protein from a mammalian cell lines for use in medicine.
	a) Purification of the recombinant product b) Introduction of genetic material into cultured cells
	c) Clinical studies d) Isolation of the gene
	e) Cloning into an expression system f) Selection of expressing cells
	g) Characterisation of the product h) Large scale culture
	i) Pre-clinical studies

10.8 Medical applications of genetic engineering

In this chapter we have concentrated on the biotechnological applications of genetic engineering. However it is becoming more and more likely that genetic engineering will soon play an important therapeutic role in medicine. Here we will briefly examine some of these applications.

10.8.1 Genetic disorders

Ultimately, it will be possible to correct genetic disorders by replacing defective genes. Certain types of genetic disease will be particularly suited to treatment by gene therapy in the near future. These are the single gene disorders where the gene is not extremely stringently regulated. Two types of gene therapy are envisaged:

- germline gene therapy;

- somatic cell therapy.

Some somatic cells are particularly suitable to act as recipients for genes. The haemopoietic system has pluripotent stem cells, and, in animal models, these cells have been infected with recombinant retroviruses carrying a human gene. The gene was found to be expressed in the peripheral blood lymphocytes for over 10 months, although expression was not very efficient.

Advances have also been made in the introduction of foreign genes into cultured skin and blood vessel cells by insertional mutagenesis.

Retroviral-mediated gene transfer is the current method of choice for clinical studies, although any treatment involving retroviruses carries a risk that the defective retrovirus may recombine with non-defective retroviruses, which would allow its spread to inappropriate sites. In 1989, the first approved human gene experiment was conducted in the United States. Tumour infiltrating lymphocytes (TILs) were engineered to carry a marker gene for neomycin resistance into tumours of patients with terminal melanoma. In 1990 the same research group transfused a small girl suffering from adenosine deaminase deficiency with lymphocytes which had been infected with a retrovirus bearing the adenosine deaminase (ADA) gene. The trial appears to have been successful, although this is obviously too small a sample size to draw any statistically valid conclusions.

Cancer treatment

Another application of gene therapy is in the treatment of localised proliferating tumours. In an animal model, rats with cerebral glioma were given an intra-tumoural injection of mouse fibroblasts that were producing a retroviral vector expressing Herpes simplex virus thymidine kinase. The thymidine kinase product turns any cell which produces it into a target for the anti-viral drug ganciclovir. After 5 days, during which the retroviruses transduced the neighbouring proliferating glioma cells, the rats were treated with the anti-viral drug. The gliomas regressed completely in 11 out of 14 rats, and in none of the controls. The use of a similar system in inoperable human brain tumours was approved by the National Institute of Health in the United States in June 1992.

Delivery of products *in vivo*

Many humans diseases could be cured if the right product could be made by a cell at the right time. One of the directions of clinical research has been to move from the production of the missing compounds by pharmaceutical companies, towards engineering patient's cells to make the products themselves. The obvious candidates for genetic engineering are those cells which can be easily taken out of the body and replaced after they have been modified *in vitro*. There has been much interest in blood and bone marrow, although these are difficult to infect with retroviral constructs, and, when these cells mature, expression of the recombinant protein stops. Some scientists have turned to endothelial cells (the cells which line the bloods vessels and airways of the lungs) which are easier to infect. After introducing the gene of interest into endothelial cells, the cells are re-introduced into the animal in the form of organoids (see Chapter 9). Such organoids have been re-implanted into the peritoneal cavity of rats, where they induced the formation of new blood vessels.

Summary and objectives

We began this chapter by indicating why animal cells are important for the large scale production of proteins especially those of therapeutic use. In many instances the ability to produce these materials on a sufficiently large scale depends upon isolating and modifying existing genes and inserting them into new host cells. The main focus of this chapter has been on how to genetically manipulate animal cells *in vitro* using recombinant DNA technology. We have described the different methods of introducing genetic material into animal cells *in vitro*. These methods generally use vectors which have been designed to introduce DNA and to correctly express the gene product in animal cells. These vectors are based on plasmids or viruses which can be engineered to contain the gene of interest, genes which allow for selection of transfected cells, regions which facilitate gene expression, and amplification signals. There are a number of different methods of introducing the engineered vectors into the cell including viral infection, calcium phosphate transfection, lipofection, microinjection, electroporation, protoplast fusion and microprojectiles from a gun. The choice of method depends mainly on the vector used and the host cell type. Once the genetic material has been introduced, there is usually a selection procedure in order to eliminate those cells which have not taken up any foreign DNA. Stable cell lines are isolated and analysed for integration of the desired gene, and expression of the gene product. The genetically engineered animal cells may be used to generate proteins of therapeutic or diagnostic value in which case a number of tests have to be performed to check the safety of the gene product.

Towards the end of the chapter, we examined the application of genetic engineering of animal cells for medical use.

Now that you have completed this chapter, you should be able to:

- list the relative advantages and disadvantages of using mammalian and bacterial cell culture for the production of protein;

- outline the steps used to transfer desired genes from various sources into animal cells;

- describe the important elements which control the expression of structural genes in mammals;

- draw examples of some vectors that can be used to transfect animal cells and explain their major features;

- outline the process by which retroviral gene sequences become integrated into their host's genome;

- list the techniques that may be used to introduce genetic material into animal cells in culture;

- make informed decisions about the choice of vector and host system;

- explain how stable, transfected, cell lines may be isolated;

- identify factors that may influence the extent of expression of a gene of interest in a host cell;

- explain how cell lines may be analysed for the incorporation and expression of 'genes of interest'.

Responses to SAQs

Responses to Chapter 1 SAQs

1.1   There are several ways in which the application of tissue cultures can be categorised. We suggest the following three categories

1) Use of the cells themselves as a product for grafting purposes. Applications include skin cell growth *in vitro* which is already being used in the treatment of burn wounds.

2) The cells can be used in the production of vaccines.

3) The cells can be used to produce proteins, for example proteins destined for pharmaceutical use.

Obviously there are many other uses of cultured animal cells in research, but these categories cover the important products which are made by cultured animal cells. Do not worry if you used slightly different categories.

1.2   1) True. Normal epithelial cells are anchorage-dependent and will only grow and divide *in vitro* if they are attached to a substratum.

2) False. Non-malignant human cells show a finite life *in vitro* and after several cell divisions, they die. They are, therefore, mortal and not immortal.

3) True. Cells from adults are more highly differentiated and are inherently more difficult to cultivate *in vitro*.

4) True. Cells derived from haemopoietic stem cells are blood cells. Typically these cells are relatively non-adhesive (think about the role of blood cells, they need to circulate. If they were adhesive then they would not fullfill their function). They usually show little or no anchorage-dependence.

5) False. These cells will grow in suspension (that is they are anchorage-independent) and do not require a large surface to which to adhere. However, usually the vessels used are designed to have a large liquid:air interface to allow for the diffusion of O_2 required by these cells.

6) False. Although the extracellular matrix is not essential for the growth and division of cells, the matrix does more than simply bind cells together. It provides the environment in which cells develop and differentiate. Thus cells grown in the presence and absence of the extracellular matrix show many differences - a clear indication that the extracellular matric has a role to play in cell differentiation.

1.3   Your arrow should have been placed somewhere along the shaded portion.

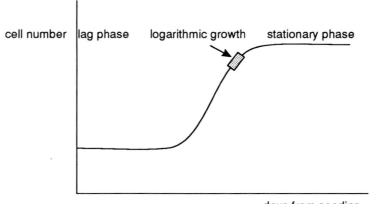

If we passaged the cells too early, we will have a too low cell density in the sub-culture. If we passage the cells too late, many of the cells will have stopped growing and might have died.

1.4

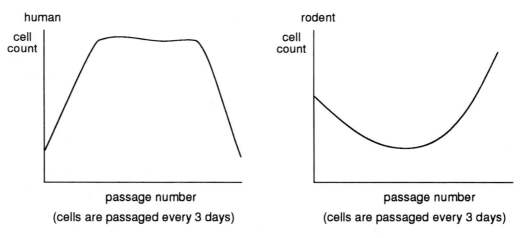

1.5   You may have listed any three of the following changes for this question.

1) Different growth rate. The cells which have undergone crisis tend to have higher growth rates.

2) Reduced size of cells which have undergone crisis.

3) Higher cloning efficiency.

4) Increased tumorigenicity.

5) Variable chromosome complement.

1.6   We suggest you should have placed the studies in the following categories.

Organ culture:

Histological studies

Analysis of the mechanism of differentiation

Analysis of cell-cell interactions

Analysis of cell-extracellular matrix interactions

Cell culture:

Replicate sampling and quantitation

Propagation of large numbers of identical cells for analysis

Studies on DNA synthesis

Analysis of cell-extracellular matrix interactions[*]

Drug toxicity analysis

Production of monoclonal antibodies

Clearly the studies which involved intercellular interactions or interactions between cells and an extracellular matrix need to be undertaken with cells that are associated with each other (ie in organ culture). Many of the studies listed under cell culture could also be studied using organ culture but would not be so easily studied or provide such reproducible data using the latter type of culture technique.

[*]Note that we have included the analysis of cell-extracellular matrix interactions under both organ and cell culture. Both can be used to provide insights into these interactions.

Responses to Chapter 2 SAQs

2.1   Enzyme solutions cannot be sterilised by autoclaving since they would be denatured. The usual way of sterilising such solutions is to pass them through sterile, fine pored, membrane filters. Typically these have pores with 0.22 μm diameters. These are sufficiently fine to remove all bacteria and larger microbes. They do not, however, guarantee the removal of viruses. If you want to ensure that the enzyme preparation is virus-free, much finer filters are required. The problem with these is that they have slow flow rates and soon become clogged. This, however, is not usually a problem if only small volumes of enzyme solution are to be filtered.

2.2   Collagen is restricted to the extracellular matrix and thus the action of collagenase is limited to changes in this matrix. Trypsin hydrolyses proteins both in the extracellular matrix and on the surface of cells. It tends to be more damaging. You should, however, not think that no damage is accrued to cells through the use of collagenase. Disruption of the extracellular matrix inevitably has some consequences on the environment to which the cells are exposed and this may lead to some cell disfunction.

2.3   You may not know, without further information, what Hu B stands for. You may have guessed that the cells were of human origin (Hu) and you may have assumed that B stands for B cells. (You would need to confirm this with your supplier).

The fact that the culture has been designated clone 1A2 indicates that this culture had been derived from cloned cells. The /5 indicates that it has passed through 5 generations (sub-cultures).

2.4 2.83 x 10^6 since cell density $= \dfrac{283}{0.1 \times 10^{-3}}$ ml^{-1}

2.5

1) We would anticipate that the macrophages would be mainly found in a) since they are very adhesive cells. T cells are, on the other hand relatively non-adhesive and we would expect to find them mainly in d).

2) No. The process described in the question is not 100% efficient. Thus although a) will contain mainly macrophages, other cell types may also be present. We are unlikely to produce an absolutely pure clone by a single passage through this scheme.

Responses to Chapter 3 SAQs

3.1

1) Undefined. We do not know exactly what is in the serum, therefore, we cannot define the exact composition of the medium.

2) There are several alternatives. A bicarbonate:CO_2 buffer might be appropriate since it will buffer at this pH. You may have considered using other buffers such as HEPES (N-2-hydroxyethyl piperazine-N-ethanosulphonic acid) as this will also buffer at this pH. Bicarbonate-based buffers are particularly useful for cultivating cells incubated in air containing 5% (v/v) CO_2. We will discuss this in the text in more detail later.

3) No. Phosphates will be needed by the cells to produce phosphorylated compounds such as nucleic acids and phospholipids. Therefore, we will have to include some phosphate(s) in the medium.

3.2 You should have concluded that the answer is probably no. Serum is obtained from animals and the exact composition of the serum depends on the animal (age, physiological status, feeding). There is variability between animals (for example in the concentration of growth factors, hormones in the blood stream). We might anticipate that different batches of sera are of slightly different composition and that these differences may be reflected by the extent to which they will support growth. Manufacturers attempt to unify their product as much as possible but it must be anticipated that there would be some variations.

3.3 Inclusion of phenol red as a pH indicator means that we have an *in situ* pH monitoring device. This means we do not have to keep opening the vessel to take out portions to measure the pH, or to open the vessel to insert a pH probe to measure the pH. Thus this reduces the chances of contamination. Furthermore, phenol red can be used in cultures of all sizes. Its disadvantages are that the pH indicator may be toxic to cells and it does not give pH values over a continuous scale - it only tells us if the pH is above or below the pKa value of the dye. Fortunately, the presence of serum has been shown to reduce the toxicity of this indicator.

3.4

1) Probably the calcium chloride would supply so much Ca^{2+} ions that these would begin to precipitate out with the phosphate. On autoclaving, the HCO_3^- ions would be converted to CO_3^{2-} ions ($2HCO_3^- \rightarrow H_2O + CO_2 + CO_3^{2-}$) which would precipitate out the calcium ions as the insoluble carbonate. Autoclaving would also cause the glucose to caramalise.

The point we are making in this question is that we have to think carefully about the formulation of a medium to ensure that it not only supplies the needs of the cells but is not subjected to undesirable side effects such as co-precipitation of ingredients.

2) The low calcium concentration indicates that this medium is used to cultivate unattached suspensions of cells. Joklik's medium is in fact especially designed to cultivate cells in suspension.

3) This is not a straight forward question. Several approaches are possible. It would be wise to select a variety of media to try. For example, we would suggest that you would try SFRE 199 medium as this has been designed to cultivate kidney cells albeit from baboons. It might work with the kidney cells from cats. Similarly we might also try Dulbecco's Modified Eagle's medium and/or NCTC 109 medium as these are both good general purpose media. RPMI and the MCDB series were designed for quite different cell lines and there is no reason to believe that they will be satisfactory for the kidney cells.

You might be tempted to supplement the selected media (a, b and j) with serum. This may improve growth yields of cells but, of course, we loose the defined nature of the media. Also we have the difficulty of choosing a particular serum since the supplier does not market cat serum. There is little logic behind choosing any particular serum from those listed. Foetal serum may be most satisfactory but there is little hard evidence on which to base such an assertion.

The point we are making with this question is that in the early stages of cultivating a particular cell type, there is rather limited logic that we can bring to bear in selecting a particular medium to use. The approach is rather empirical. It is, however, very worthwhile researching the literature thoroughly, it could save you a lot of time and frustration, especially if you find reports of previous successful cultivations of the cell type of interest.

3.5   Arguments for adding carboxymethyl cellulose before sterilisation is that is will be sterilised with the medium ingredients and this reduces the number of additions that need to be made to the medium after sterilisation. This will, therefore, reduce the chances of contamination of the medium. On the other hand, this cellulose derivative is viscous and since we may need to filter sterilise the medium, this could create problems.

In other words, their are advantages and disadvantages in sterilising the two (medium and CM-cellulose) together or separately. The choice may be dependent upon the size of the operation and the actual concentration of CM-cellulose to be used.

3.6   The correct combinations are:

a) T-cells   interleukin-2
b) Skin explant   epidermal growth factor
c) liver cells   gly-his-lys (liver cell growth factor)
d) fibroblasts   FGF (fibroblast growth factor)
e) ganglia   NGF-β (nerve growth factor β)

You might have known all of these, it depends on how much experience you have of cell biology and physiology. Nevertheless, we are confident that you have understood the point being made that particular cells require particular growth factors (or combination of growth factors).

Responses to Chapter 4 SAQs

4.1   You should have been able to detect that the banding pattern obtained with cell line C was different from that obtained with cell lines A, B and D using the mini satellite probe 1. Thus, cell line C must be different from the other cell lines. This is further confirmed in the second autoradiogram. In this second autoradiogram, however, A also appears to be different from B and D.

From these results, it would appear that B and D may be identical, but that these cell lines are different from A and C.

This question was set to illustrate the principles involved in DNA fingerprinting and to enable you to demonstrate that you can interpret data from such a procedure.

In practice, even closely related cells can give quite different banding patterns and thus the identification of cells is rather easy to establish. The difficulty with the technique is detecting subtle differences between two cell lines. Say, for example, we have two cell lines derived from the same individual. A single or few base changes in a single gene may have considerable effect on the phenotype of a cell but may go undetected by the DNA-fingerprinting technique.

Despite this uncertainty, DNA fingerprinting has the potential of becoming the standard reference technique for identifying cells.

It would appear that probe 2 routinely detects the differences between (A) and (B and D) and also the differences between (C) and (B and D). Thus it appears that this probe is satisfactory. Probe 1 does not show up any differences between (A) and (B and D). In other words, it provides no additional information to that gained from using probe 2. This indicates that if only one probe was to be used, then probe 2 should be selected. Of course, the use of more than one probe may lead to the detection of other differences.

4.2   1) A simple test would be to check which type of G6PDH is present. HeLa cells contain type B whilst many (but not all) human cells produce type A. If type A was present, it would indicate that the culture was contaminated. If type A was absent, we would not, however, have proven that the culture was not contaminated. In this case we might have to resort to DNA fingerprinting or testing for HL-A.

2) There are several possibilities (eg using isoenzyme patterns or using surface antigens). One sure way would be to use DNA fingerprinting and to compare the autoradiograms with autoradiograms produced with the original cell line.

3) We would suggest using G11-banding as this gives differentiated and distinctive staining with mouse and human chromosomes.

4) We could use HL-A characterisation as each individual produces its own characteristic HLAs. Alternatively DNA fingerprinting could be used.

5) We would suggest using DNA fingerprinting as cells of common origin should contain indentical DNA profiles.

Note that you may have suggested alternatives to those given in our responses. In many cases more than one approach can be successfully used. The actual choice, in practice, depends on such factors as the availability of equipment and expertise. For example in 5) we suggested using DNA fingerprinting. This, of course, can only be used if we have suitable DNA probes for the cell lines under investigation. The choice of method is also dependent upon how definitive/sensitive the result needs to be. For example, checking for purity of a CHO culture (see 2), might just be a routine check for gross contamination. In this case, isoenzyme patterns may be sufficient. DNA fingerprinting is, however, more sensitive but more difficult to perform. The choice in this case, will also be dependent upon the nature of the suspected contamination. For example, if the suspected contamination is by other lines of CHO cells, then these may not be detected by isoenzyme analysis. The point we are making is that the choice of method depends upon the circumstances in which the analysis is to be done.

You have, however, learnt of many different techniques from which an appropriate selection can be made.

4.3

1) a) Species verification was covered in section 4.2 and the techniques involved are cytogenetics, isoenzymology and immunological tests.

b) The identification of the individual from which a cell line was derived depends upon genome analysis using cytogenetics and recombinant DNA methodology and upon protein analysis using high resolution 2-D gel electrophoresis, allozyme analysis and antigen analysis (section 4.3).

2) a) Only analysis of protein expression could play a role in the characterisation of cell types since the genetic material in any one individual will be the same in each cell irrespective of the cell type. Thus analysis of genetic material will give no insight into cell type.

4.4

1) is matched with b). Express the milk protein casein → human mammary epithelial cells

2) is matched with b) and d). Have a bipolar morphology → mouse embryo fibroblasts + mammary fibroblasts.

3) is matched with e). Express haemoglobin → erythrocytes.

4) is matched with c) and d). Express fibronectin → mouse embryo fibroblasts + mammary fibroblasts.

5) is matched with b). Expresses a cytokeratin → human mammary epithelial cells.

6) is matched with a). Express desmin → myoblasts.

7) is matched with b). Have a regular, polygonal morphology-human mammary epithelial cells.

4.5

It would appear that culture B was contaminated. The results for cultures A and C are very similar to the negative control (control 1) in which no mycoplasma are present. It could be, however, that there is an extremely low level of contamination. Thus it might be advisable to carry out further tests either by using more cells (and therefore more mycoplasma) to see if the differences between control 1 and these cultures is significant or attempt to sub-culture from cultures A and C in a mycoplasma medium.

4.6

1) Typically you would need to screen for viruses which may infect hamster ovary cells. You would, therefore, first need to find out what viruses infect such animals. It would also be useful to find out if these viruses only infect certain cell types. If so this may enable you to delete them from your list. However, in the first instance, it would be safer to screen for all hamster viruses.

2) a) Virulent viruses may be detected by the presence of virally coded antigens (eg viral coat proteins) using immunological assays. In these, antibodies are used which specifically bind with these antigens. There are various ways of detecting such interactions but increasingly the antibodies used are linked to enzymes and the enzyme is used as a form of label (enzyme linked immuno-sorbent assays ELISA).

b) Latent viruses are more difficult to detect. If H^3-DNA probes, which have nucleotide sequences complementary to viral genes, are available these may be used in a system analogous to that employed by Gen-probe to detect mycoplasma.

Responses to Chapter 5 SAQs

5.1

As the temperature falls, the rate of metabolism is reduced. Progressively as the temperature falls from 37°C to 0°C the rate of enzyme catalysed reactions will be slowed down. Not all enzymes are affected to the same extent and metabolism may become unbalanced. If cells are held for too long at temperature between 37°C and 0°C, the metabolic unbalance may lead to loss of viability.

At the same time, the liquid crystal structure of cell membranes will solidify. They will, therefore, become rigid and brittle and liable to fracture.

At temperatures below 0°C, the water will begin to crystallise (form ice). Thus the milieu within the cell will no longer be fluid and reactions will slow down almost (but not entirely) to a halt. You should note that the water of hydration around proteins and nucleic acids may remain fluid at temperatures well below 0°C. At temperatures below 0°C, the cells become solid as though filled with small sharp edged grains. Thus these cells need to be handled gently to prevent physical abrasion.

At even lower temperatures, the metabolism of the cells completely stops. The cells are in suspended animation rather like a dormant state. Note that even at temperatures as low as -20°C chemical changes continue to occur in cells. Thus at -20°C, the cells undergo slow chemical damage. It is only at temperatures as low as -196°C that metabolism and chemical changes are so low as to be negligible.

5.2

1) The answer is no. The viability is rather low. We usually use cultures which show a viability of greater than 95%.

5.3

1) About 15 ml of fresh media. From the viability data, there are 8×10^6 viable cells in the 5 ml of suspension, that is a cell density of 1.6×10^6 viable cells ml^{-1}.

For anchorage-independent cells, we normally use about $3\text{-}5 \times 10^5$ viable cells ml^{-1}. Thus we could add about 15 ml of medium to the cell suspension to bring its concentration to 4×10^5 viable cells ml^{-1}.

2) The % viability is very low so the method is not really very satisfactory. The question arises whether or not the cells are being killed by the freezing procedure, storage conditions or during the resuscitation process. It would be worthwhile investigating the use of alternative cryoprotectants and freezing regimes to see if viability could be improved.

5.4   There are two main reasons for depositing a cell line with a cell culture collection. Firstly, it makes the cell line available to the scientific community at large. Secondly, it means that there is now a second source of the cell line. If you have an unforseen catastrophe and lose the cell line, you can always recover the culture from the cell culture collection. Furthermore, tests carried out by the cell culture collection serve as an independent verification of the characteristics you have assigned to the cell line.

The reasons for not submitting a cell line immediately you have isolated it are:

- the cell line may not be stable. It is better to allow it to pass through several sub-cultures to prove its stability;

- you may wish to subsequently patent processes using the newly isolated cell line. If this has been previously submitted to a culture collection with unrestricted access, then you may well jeopardise your patent position.

5.5   1) The main factor which will influence your choice will be geographical location. If you are working on mainland Europe, requesting a culture from Australia or the USA means considerable transport and a threat of damage to the culture during transport. This is not the only consideration. You would also need to consider the relative cost, speed (guarantees) of service and delivery, back-up supply of information and import/export restrictions.

2) The main steps are to check viability by resuscitating the culture, check its characteristics match with the specification of the culture supplied by the cell culture collection and check for contamination.

Responses to Chapter 6 SAQs

6.1   You could have choosen focus formation, growth factor requirement and tumorigencity. The anchorage dependency assay would work with non transformed cells as well since the cells cultured from blood are anchorage independent even before they are transformed. This particular criterion of transformation is meaningless with these cells. Morphology would also be a poor choice as these cells are more or less rounded up even in the untransformed state.

6.2   This RNA virus would, within the cell, be first converted to its DNA form and then might become incorporated into the host cell's genome. There, the gene coding for the epidermal growth factor receptor would be expressed and the receptor incorporated in the surface of the cell. Such a cell would now be potentially under the control of the epidermal growth factor as well as subject to the controls which normally regulate its growth and division. In other words, we might anticipate that the growth of such a cell will be stimulated by epidermal growth factor and that it, therefore, would, to some extent, have escaped from its normal regulatory controls.

6.3   1) Exposure to high levels of sunlight (which contains UV radiation), causes thymine dimers to be produced in the DNA within the surface cells (ie skin epithelial cells). Subsequently, DNA repair mechanisms replace these dimers. If the appropriate nucleotides are inserted, then the cell will be restored to its original form. If, however, the incorrect nucleotides are inserted, the cell effectively carries mutations. If such mutations occur in regulatory sequences which control genes whose products regulate cell growth and division, then such cells may acquire new growth characteristics. Similarly, if such mutations occur in structural genes whose products regulate cell growth, these cells may also acquire changed growth characteristics. In principle, there is a probability of the cells becoming transformed if mutations occur in any of the following genes or nucleotide sequences:

- regulatory sequences controlling proto-oncogenes and tumour suppressor genes;
- structural genes coding for growth factors, growth factor receptor genes, intracellular signal transducing genes (ie proto-oncogenes);
- structural genes coding for tumour suppressors.

These mutagenic effects are mainly confined to the skin since UV only penetrates the outer surface of the body and will mutate only the cells in the outer layers.

2) The sort of mechanism we hoped you might identify is as follows. If the UV light induced a mutation of the regulatory sequence which controls a gene coding for a growth factor, the mutated epithelial cell might begin to produce the growth factor. Release of this growth factor may stimulate the growth and proliferation of the target cell type. Although such a mechanism could produce altered growth patterns in other tissues/organs, you might realise that this, if it does occur, is of very rare occurence. In fact, such a mutation would not transform the skin epithelial cell. The production of an abnormal growth factor by a single epithelial cell (or a few progeny cells derived from such a mutant cell) is unlikely to significantly alter the total growth factor level in the body and thus such mutations are likely to go undetected.

6.4 The answer is that, in principle, this might work. If the retinas of the individuals were exposed to high doses of the virus, then many of the retina cells might become infected by the virus. Being non-virulent, the virus would not proliferate but become incorporated into the host cells genomes. The expression of the Rb-1 genes in these cells would suppress the development of tumours. In practice, however, there are many difficulties in using this approach. First of all there are the practical difficulties of delivering the viruses to the retina cells and ensuring a good infection rate. Furthermore, questions arise about the safety of such a strategy. What, for example, if such viruses 'reverted' into a virulent form? There are also questions about the long term stability of the latent virus within the host cell. Will they be excised and lost?

There are also many ethical issues that need to be addressed.

Thus our answer is a rather cautious yes, it could work. We set this question to illustrate how knowledge of the molecular basis of tumor induction and development might enable biotechnologists to contribute to strategies for treating/curing the condition.

6.5 You might have generated several ideas here. Perhaps one of the most productive ideas is to consider using monoclonal antibodies. If the cancer cell produces a unique surface component (antigen) then monoclonal antibodies which bind this surface component would be specific for these cancer cells. Thus if we linked the vinblastine to such an antibody, it might be delivered to the target cell. In practice there are many difficulties with this approach. The key issues are to find cancer specific antigens and to find a mechanism for releasing the drug at the target site and not before it arrives there.

Again we set this question to show that biotechnology is not only providing knowledge of how cancers are induced but how this technology might also be employed to developing strategies for treating/curing these conditions.

Responses to Chapter 7 SAQs

7.1 There is no straight forward answer to this question. Options 1 and 3 could both be satisfactory but both have problems. Option 1 would produce an antiserum containing many different antibodies (it would be polyclonal). These would react with the antigen present on the viruses used to produce it. It is likely that some of the antigens are shared by all strains of Hepatitis B viruses and, therefore, this serum would detect all Hepatitis B infections. On the other hand, such a serum may contain antibodies which react with other viruses and would, therefore, give false positive results.

Option 3 would be more specific and would only react with antigens carrying epitope a. If, as seems probable, all Hepatitis B viruses carry this epitope, such a monoclonal preparation could be used to detect all Hepatitis B infections. This approach is based, however, on the assumption that all Hepatitis B strains produce the a epitope.

Option 2 would not be satisfactory. Since it only contains antibodies which react specifically with the 'w' epitope, it would not detect infections of viruses which did not produce this epitope (for example the adr subtype).

Perhaps the most sensible approach is to try both types of antibody preparations. Option 1 could be used to provide a general screen for Hepatitis B infections and then positive samples could be confirmed using monoclonal antibodies. This would not only confirm Hepatitis B infections, but also allow identification of the sub-strain.

7.2 Spleens contain large amounts of lymphoid tissues and have, therefore, many resident B cells. Thus, if we collect spleens, separate the cells (by for example panning), we can harvest many B cells. There is, therefore, a high probability that such a population will contain the desirable B cell type (that is, B cells which produce the desired antibody). Alternatives to spleen would be lymph nodes but these are smaller and will be more difficult to handle. Organs such as heart and kidneys contain much lower populations of lymphocytes and offer less likelihood of being useful sources of the desired B cell (or plasma cell) type and are, therefore, rarely used.

7.3
1) Hapten. It cannot by itself stimulate the production of antibodies but does so when attached to a carrier. It is also antigenic since it will react with antibodies.

2) We could use the terms immunogen or antigen.

3) It contains a single epitope since it reacts with only a single type of antibody.

4) It carries many epitopes as it will react with many different types of antibodies.

7.4 This is a fairly open ended question but we would suggest the following. Small amounts of ^{3}H-oestrogen could be added to the aliquots of the culture supernatants and incubated for a short time (for example 1 hour). If the culture fluid contained anti-oestrogen antibodies then these will bind some (or all) of the ^{3}H-oestrogen. If, however, no such antibodies were present, the ^{3}H-oestrogen would remain free in solution.

Our task then would be to distinguish between bound and unbound ^{3}H-oestrogen. This could be done on the basis of size. In this case a simple approach to this could be to add activated charcoal. This will bind small molecules (for example ^{3}H-oestrogen) but not large molecules (for example ^{3}H-oestrogen-antibody complex). Thus if we centrifuge such a mixture, we would be left with ^{3}H-oestrogen-antibody complexes in solution. We could measure this by determining the amount of radio-activity left in solution.

We can represent this process in the following way:

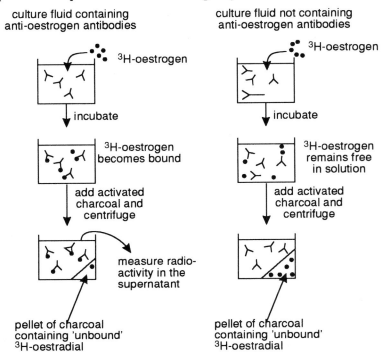

This is rather an over simplification and we need to carry out suitable controls but at least you should have understood the underpinning principle. There are many variations to this general strategy and these depend upon, for example, the nature of the antigen being used.

7.5   10

Since 0.1% of the B cells used produce an appropriate antibody, we can anticipate that 0.1% of the viable hybridomas will also produce the appropriate antibody. But, if we started with 10^{10} B cells, only 10^8 B cells will successfully fuse with the myeloma cells (ie 1%). Of these only 0.01% are viable. In other words only $10^8 \times 10^{-4} = 10^4$ viable hybridoma cell lines would be produced. Thus since we are anticipating that only 0.1% of the viable hybridomas will produce the appropriate antibodies, we might anticipate that we may produce 10 such hybridomas.

Responses to Chapter 8 SAQs

8.1   1) The data show that over the first 40 hours of incubation, the rate of glucose consumption as measured by the change in glucose concentration in the outflow increases. This indicates that the biomass concentration in the reactor is progressively increasing. After 50 hours, the rate of glucose consumption declines markedly. This can be explained in the following way. Up to 50 hours, the glucose is used to support growth. However, after 50 hours there is confluent growth of the cells over the surface of the beads so the cells no longer grow (ie they are contact inhibited). At this stage, the glucose that is consumed is to provide energy for maintenance purposes (eg repair, maintenance of osmotic balance etc).

Thus we may conclude that over the period 2-40 hours there is extensive cell growth (possible logarithmically). But after about 40 hours this growth rate slows down and eventually (by about 60 hours) stops. At this stage, the cells are in stationary phase. They nevertheless still have a requirement for energy.

If we continually monitored glucose consumption, measured by ($[Glucose]_{in}$ - $[Glucose]_{out}$) x flow rate, we would expect to produce a relationship like this:

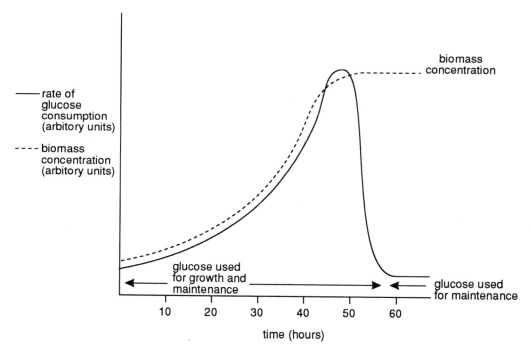

2) In this experiment, the glucose concentration in the outflow remains more-or-less constant throughout the incubation. Thus glucose utilisation remains more-or-less constant. You might have thought that this means that the cells are not growing. This is not necessarily so. The results do, however, indicate that the culture is glucose limited. The limited glucose supply means that there is a competition for this substrate between supporting growth and fulfilling maintenance functions.

Thus:

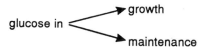

Note that some glucose remains in the culture medium. The amount left depends on the efficiency of uptake by the cells' glucose transport mechanism. If this was avid, then the glucose concentration in the outflow would be low. The point we are making in this SAQ is that the rate of substrate consumption is not just dependent upon the number of cells present, but depends upon the activities of these cells. If substrate is limiting, then the rate of substrate consumption is not a measure of the amount of biomass present.

Another way of expressing this is to consider the relationship we gave in the text.

$$\text{specific substrate utilisation rate} = \frac{\text{change in substrate concentration (d[S])}}{\text{time (dt) x cell mass or number ([X])}}$$

The specific substrate utilisation rate (ie rate of substrate use per unit of biomass) dependes on the metabolic activities of the cell. The relationship only holds when substrate is not rate limiting.

There is no evidence that the culture described in 2) is in the logarithmic growth phase. There is, infact, little evidence of any increase in biomass over the incubation period.

8.2   In the initial culture vessel, the rate of O_2 transfer = 17 000 µg h^{-1} (surface area x transfer rate)

This is more than adequate to supply the oxygen to this culture.

In the larger vessel, we estimate that the rate of oxygen consumption will be 5000 x 10 µg h^{-1} = 50 000 µg h^{-1} (remember we are using 10 times larger volume of culture).

The rate of oxygen transfer in this system will be 200 x 170 µg h^{-1} = 34 000 µg h^{-1}. In other words, this system will be unable to supply all the oxygen that is required when the culture reaches the late log phase.

There are three approaches to solving this. One is to increase the concentration of oxygen in the air supplied to the vessel. This will increase the oxygen transfer rate.

The second approach is to sparge air through the culture.

The third is to increase stirrer rate.

In practice, the actual choice will depend upon the ability of the cells to withstand the shear forces, the relative costs of sparging, increasing oxygen tenson or increasing stirrer speeds and the availability of equipment.

8.3   1) Specific growth rate will be 1.6 day^{-1}, and the yield will be 8 x 10^9 cells day^{-1}

The specific growth rate will be equal to the dilution rate since $\mu = D$ when a steady state has been reached.

Since $D = \dfrac{F}{V} = \dfrac{8}{5} = 1.6$ day^{-1}, then $\mu = 1.6$ day^{-1}

The yield will be F x [X] = 8 x 1 x 10^6 x 1000 (note cell density was given in ml^{-1}) = 8 x 10^9 cells day^{-1}

2) This probably reflects that D (which is now = $\dfrac{20}{5}$ = 4 day^{-1}) is higher than μ_{max} for these cells. Therefore, the cells are being 'washed out' of the vessel before they can be replaced by growth and division.

8.4   1) This applies to spinner culture, batch stirred tank reactors, air-lift reactors and continuous flow stirred tank reactors.

2) Typically applies to batch systems (spinner cultures, batch stirred tank reactors and air-lift reactors). Would also apply to membrane reactors used in batch format.

3) This is a consequence of producing low density culture and, therefore, you should have selected the reactors listed in 1).

4) Typically stirred tank and spinner cultures may produce high shear forces. This is also true of air-lift reactors if high air throughputs are used.

5) Membrane/hollow fibre reactors; also stirred tank reactors used in continuous format should be included.

6) All the vessels are susceptible to contamination. However, systems which require pumping media or gases into vessels increases the risk of contamination. Thus we would anticipate that you would have suggested continuous flow stirred tank reactors and membrane reactors.

7) Membrane/hollow fibre reactors operated in continuous mode and continuous flow stirred tank reactors. With continuous systems, we can establish steady state (ie constant) conditions within the reactor.

8) Membrane/hollow fibre reactors and continuous flow stirred tank reactors. All batch systems have to be periodically emptied, cleaned and recharged. They, therefore, have extensive 'down times'.

8.5 Up to 10^5 individuals.

Since you calculate that you can produce $50 \times 2 \times 10^3$ litres of culture per annum, then you can produce $\dfrac{50 \times 2 \times 10^3}{0.1}$ doses of interferon

$= 1 \times 10^6$ doses of interferon.

But individuals require 10 doses, so in principle you can produce sufficient interferon for 10^5 individuals. In practice, however, some of your material would have to be used for quality control and standardisation. Thus the actual number of individuals that could be supplied would be somewhat less than 10^5.

Responses to Chapter 9 SAQs

9.1 1) Monolayer culture of a pure cell strain on plastic = cell culture.

2) Monolayer culture of a pure cell strain on collagen = cell culture.

3) 3-D culture using one pure cell strain grown at high density = histotypic culture.

4) 3-D culture using more than one cell strain = organotypic culture.

5) 3-D culture using more than one cell strain and 3-D matrices (eg sponges/gels) = organotypic culture.

6) 3-D cultivation of tissue so as to retain its architecture = organ culture.

9.2 Do not worry if you find this question difficult. We included it because it makes the point that it is not always easy to categorise different types of tissue culture as we have done so far.

1) Growth of a pure cell strain on a feeder layer of a different cell type.

Culturing cells on top of a feeder layer of cells can enhance growth and/or differentiation of those cells (section 4.4.4). This could be described as cell culture or histotypic culture, and the choice of description used probably depends to a large extent upon what the cells do under these conditions. In other words, whether they form three-dimensional structures which resembles the native tissue or whether they grow as monolayers.

2) Recombination of different tissue types (from tissue not from dispersed cells).

This is strictly speaking organ culture, since the starting material is not a dispersed cell line or strain, and the tissue structure is still retained, although some reconstruction of the tissue is obviously involved. Recombinations of this type were used in the classic experiments performed by Grobstein in 1954. He demonstrated the role of cell-cell interactions in embryonic induction using the mammary gland as a model. Mammary epithelial cells recombined with mammary stroma to develop a typical mammary pattern whereas if they were combined with salivary stroma their morphogenesis was that of a salivary gland. His studies demonstrated that embryonic determination is largely a consequence of stromal influences. Furthermore, his studies led to the finding that it is the accumulation of extracellular matrix between the two tissues which facilitates this stage of development.

3) Monolayer culture of primary cell culture.

A primary explant could be used in cell culture or in organ culture, depending on which type of growth is encouraged.

4) Whole embryo culture.

This is closest to organ culture although, since it is the cultivation of a whole organism, it represents a separate category of tissue culture which will be described presently.

9.3   The main constants are that the embryo and recipients must be compatible. Usually this means, of the same species. In addition the uterus must be in the right state to receive the embryo. Thus the recipient cannot be at just any stage of the oestrous cycle. Usually they are brought to the right state (ie just post-ovulation and well prior to menstruation) using hormone treatment.

9.4   1) Invasive. The dye used has to permeate between the cells of the embryo.

2) Non-invasive. This can be done by microscopic observation.

3) Non-invasive. This can be done by microscopic observation.

4) Invasive. Although the DNA probe is used on a cell (or cells) outside of the embryo, the embryo has to be 'invaded' to remove the cell.

5) Invasive. The embryo has to be immersed in a solution containing the antibody which adheres to the surface of the embryonic cells.

6) Non-invasive. Nutrient uptake can be determined using samples of the culture medium.

Responses to Chapter 10 SAQs

10.1   There are many advantages and disadvantages in using bacterial and mammalian cell cultures. Below we have cited some of the main ones. You may have thought of many others.

	Advantages	**Disadvantages**
Bacterial cell culture	Bacterial cells are easier to work with than mammalian cells. Rapid growth/simple (cheap) culture media. Genetically simpler to manipulate, more methods available to transfer genes. More known about bacterial genomes and gene regulation	Products differ from products of mammalian cells -for example in glycosylation pattern, but also in proteolytic cleavage, phosphorylation and other post-translational modifications
	Bacterial cells can produce more protein than mammalian cells	Foreign proteins expressed in bacteria often form clumps of denatured proteins or alternatively, the foreign protein may be broken down.
	Fewer ethical issues	Some bacteria are pathogenic
Mammalian cell culture	Proteins can be processed correctly by mammalian cells	Animal cells are more fragile than bacterial cells
	Mammalian cells can be engineered to secrete their products into the cell culture medium, making purification easier	Animal cells have slower growth rates than bacterial cells
		Cell densities are usually lower
		Animal cell cultures have more problems with contamination than bacterial cell cultures (result of richer media/longer cultivation time/faster growth of bacteria)
		Ethical concerns may be greater

10.2

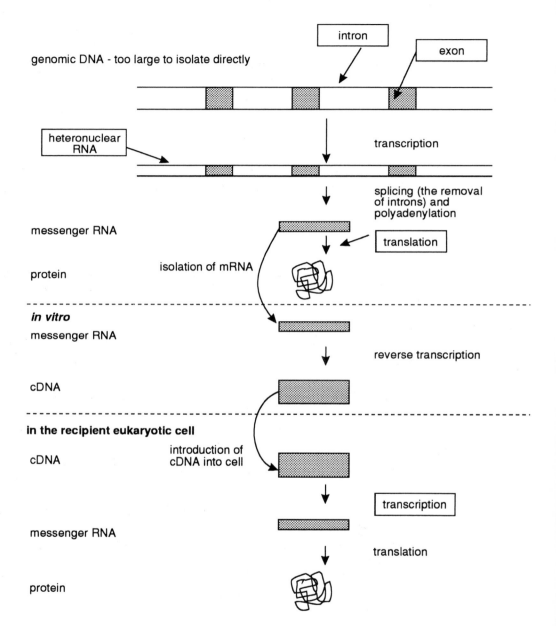

10.3 The answer to this question is not immediately obvious but it lies in the fact that even when the final aim is to introduce the recombinant plasmid into eukaryotic cells, the plasmid needs to be able to replicate in bacteria. This is because only bacteria produce sufficient amounts of plasmid. The ampicillin gene facilitates the selection of bacteria which contain the recombinant plasmid.

10.4 a) Will replicate in mammalian cells providing the host cells produce the polyoma virus T antigen.

b) Will replicate in suitable mammalian cells as it has a suitable *ori* gene.

c) Will replicate in mammalian cells since it has a suitable origin (SV40 *ori*) and a gene coding for the SV40 T antigen (assuming that this gene is expressed).

d) Will only replicate in mammalian cells if the cells are expressing the appropriate T antigen (ie SV40 T antigen).

10.5   The important control elements in a typical mammalian protein gene.

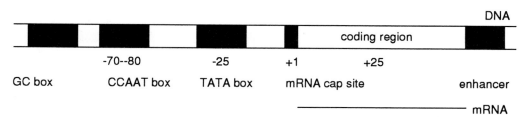

Well done if you got the enhancer right. This was a tricky one, but remember that enhancers can be on either side of the gene, whereas the promoter elements are found at the 5′ end of the gene.

10.6

1) The SV40 early region. This is a viral promoter region which is frequently used in mammalian expression systems to obtain high levels of constitutive gene expression.

2) Polyadenylation signals. These need to be added when cDNA is used, so that a poly A tail can be added to the newly transcribed mRNA.

3) Promoter. This often needs to be added to a gene construct, especially when cDNA is used, in order to obtain expression.

4) Proteins. Proteins are not genetic elements so this term does not fit in the list.

5) The Shine-Dalgarno sequence. This is a prokaryotic genetic element which is involved in ribosome recognition. It would not be used in gene expression in animal cells.

6) The recognition sequence for translation. Yes, again this needs to be present in the gene construct.

7) Secretion signals. These are optional extras. Where large amounts of a product are needed, it may be more efficient to secrete the protein into the medium and collect the medium rather than to collect cells and extract the desired product from them.

8) The Pribnow box. This is a prokaryotic genetic element and would not be used in animal expression systems.

9) Enhancers. These may be required in addition to promoters, and are especially important in tissue specific expression.

10) cDNA. The cDNA may be used as the source of the actual genetic material, in other words it contains the information needed to make the product.

11) *dhfr*. This may be used where gene amplification is desired.

12) *amp*^{R}. Although this is a bacterial gene, it is usually included in constructs used for expression in animal cells since there is generally an amplification stage which takes

place in bacteria. The amp^R provides a means of selecting bacterial cells that have received foreign DNA.

10.7

1) amp^R provides a means of selecting bacterial cells that have received foreign DNA.

2) neo^R provides a means of selecting eukaryotic cells that have received foreign DNA.

3) SV40 early region facilitates constitutive expression of genes in eukaryotes.

4) LTR facilitates inducible expression of genes in eukaryotes.

5) *dhfr* allows for amplication of genes.

10.8

1) The likely outcome would be that the viral RNA and proteins (coat/envelope) would be produced but would fail to be packaged properly. Thus there would be a build up of viral components and poorly structured particles within the cell.

2) It is likely that all the viral components, except the *gag* proteins, (nucleoproteins, core proteins, capsid protein) would be produced and accumulate in the cell. Intact viral particles would not be produced.

3) It is probable that the retrovirus would be replicated. The *pol* gene products coded for by the intact virus would also synthesise DNA from the defective virus. The defective virus on its own would not be replicated since there would be no mechanism available for its RNA sequence to be transcribed into a DNA sequence.

4) Again it is likely that intact viruses would be produced but in this case the virus would be coated by *env* proteins coded for by the intact retrovirus. There are, however, some limits on how successfully the 'wrong' envelope proteins can coat viral core particles.

10.9   Your flow chart should look similar to our figure below.

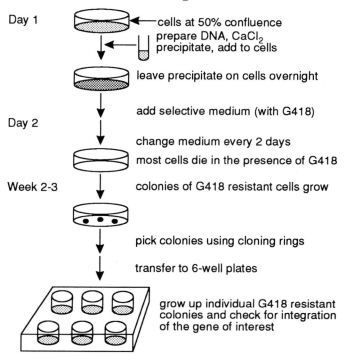

10.10   We have mentioned many different cells lines in this book. Do not worry if you did not
remember all of them, or if you remembered different interesting points about them.
Here are the many cell lines we have mentioned.

Cell line	Why appropriate
Xenopus oocytes	Useful for micro-injection of DNA or RNA since the cells are very large
Sf9 cells	Derived from the fall armyworm and used for baculovirus infection
COS cells	African green monkey kidney cells which produce the SV40 large T antigen. Used with plasmid harbouring SV40 origin of replication
WOP cells	3T3 cells which express polyoma virus T antigen. Used with expression vectors containing polyoma origin of replication
ES-cells	Can be genetically manipulated. Can be used to create a whole organism carrying the recombinant genes

CHO cells	Chinese hamster ovary cells, *dhfr* deficient CHO cell lines exist and are used in the amplification strategy (section 10.2.7), used to make vaccines against influenza virus and HIV
3T3 cell lines	Widely used rodent embryonic cell lines, very useful for analysis of oncogenes (see Chapter 6)
Myeloma cells	Malignant counterparts of immunoglobulin secreting plasma cells, particularly useful in the production of immunoglobulins.

10.11   There are several possible reasons why this result was observed.

Firstly, the evidence from the Southern blotting procedure indicate that the cell contains the gene.

The failure to detect the corresponding RNA molecules by the Northern blotting procedure could reflect the fact that the gene was not being expressed or alternatively the pool of mRNA in the cell derived from this gene was so small that the method was not sensitive enough to detect it. A third possibility was that the mRNA was degraded by RNase during extraction and isolation.

10.12   1) Integration site - yes.

2) Origin of gene - no.

3) Copy number - yes.

4) Orientation of enhancer - no.

5) Transcriptional efficiency - yes.

6) Number of chromosomes in host genome - no.

7) RNA stability - yes.

8) RNA processing - yes.

10.13   Your sequence should be:

d) Isolation of the gene.

e) Cloning into an expression system.

b) Introduction of genetic material into cultured cells.

f) Selection of expressing cells.

h) Large scale culture.

a) Purification of the recombinant product.

g) Characterisation of the product.

i) Pre-clinical studies.

c) Clinical studies.

Suggestions for further reading

Related texts from the BIOTOL series

Underpinning cell biology

Infrastructure and Activities of Cells, Butterworth-Heinemann, Oxford, ISBN 0-7506-1500-1

Functional Physiology, Butterworth-Heinemann, Oxford, ISBN 0-7506-0563-4

Process technology revelant to animal cell cultivation

Bioprocess Technology: Modelling and Transport Phenomena, Butterworth-Heinemann, Oxford, ISBN 0-7506-1507-9

Operational Modes of Bioreactors, Butterworth-Heinemann, Oxford, ISBN 0-7506-1508-7

Bioreactor Design and Product Yield, Butterworth-Heinemann, Oxford, ISBN 0-7506-1509-5

Genetic aspects

Genome Management in Eukaryotes, Butterworth-Heinemann, Oxford, ISBN 0-7506-0558-8

Techniques for Engineering Genes, Butterworth-Heinemann, Oxford, ISBN 0-7506-0556-1

Strategies for Engineering Organisms, Butterworth-Heinemann, Oxford, ISBN 0-7506-0559-6

Applications

Biotechnological Innovations in Animal Productivity, Butterworth-Heinemann, Oxford, ISBN 0-7506-1511-7

Biotechnological Innovations in Health Care, Butterworth-Heinemann, Oxford, ISBN 0-7506-1493-5

Technological Application of Immunochemicals, Butterworth-Heinemann, Oxford, ISBN 0-7506-0508-1

Other literature

Journals giving relevant overviews

Current Opinions in Biotechnology

Trends in Biotechnology

Trends in Cell Biology

Annual Reviews of Cell Biology

Annual Reviews of Biochemistry

Annual Reviews of Immunology

Texts

Morgan, S.I. Animal Cell Culture, 1993, Bio Scientific Publishers Ltd,, Oxford, ISBN 1-8727-4816-3

Pollard, J.W. and Walker, J.M. Animal Cell Culture, 1990, Humana Press, Totowa, NJ, USA, ISBN 0-89603-15-0

Freshney, R.I. Animal Cell Culture: A Practical Approach, IRL Press, Oxford (1986), ISBN 0-947946-33-0

Butler, M. Mammamial Cell Biotechnology: A Practical Approach (1991), IRL Press, Oxford, ISBN 0-19-963209-X

Index

A

acceptor sequences, 223
acetylation, 207
Acholesplasma, 95
acrylic polymer, 161
actin, 82
activation of proto-oncogenes, 125
adenosine, 66
adenosine deaminase (ADA) gene, 235
adenovirus 5, 228
adenovirus
     major late promoter, 217
adenylate kinase (AK-1), 91
adherence of cells, 65
adherent cultures, 9
adhesion factors, 61
adipose tissue, 82 , 88
adoptive immunotherapy, 182
ADP ribosylation, 207
adrenal, 4 , 182
adrenal medulla, 6
adult lung fibroblasts, 13
adult mammalian tissues, 187
adult rabbit liver, 187
adult tissue, 9 , 25
African Green Monkey kidney, 213
agar clot technique, 187
agarose cassette system, 77
agriculture, 7
AIDS, 6 , 208
air flow cabinet, 23 , 24
air-lift reactors, 178
albumin, 69
alkaline, 63
alleles, 80
allelic isoenzymes, 80
allozyme analysis, 80
allozymes, 80
American Type Culture Collection, 147
amino acids, 47 , 58 , 60 , 66
aminopterin, 59 , 144
ammonia, 63
amniocentesis, 7
amnion, 7
amniotic fluid, 49
amphibia, 60
amphibian oocytes, 229
amphotericin B, 94
ampicillin resistance gene, 214
amplification systems, 213
*amp*R, 212 , 214

ampullary region of oviducts, 192
analysis of isoenzymes, 90
anchorage-dependent cell lines, 9 , 120 , 167
anchorage-dependent cells, 157
    scale up of, 157
aneuploid, 14
animal blood serum, 3
animal breeding, 199
animal cell culture
    large scale, 154 , 155 , 157 , 159 , 161 , 163 , 165 ,
167 , 169 , 171 , 173 , 175 , 177 , 179 , 181 , 183
animal cell culture media, 46 , 47 , 49 , 51 , 53 , 55 ,
57 , 59 , 61 , 63 , 65 , 67 , 69 , 71
animal cells, 5 , 207
animal cells in suspension, 178
animal tissue culture in perspective, 18
animal tissues
    blood, 82
    connective, 82
    epithelial, 82
    lymph, 82
    muscular, 82
    nervous, 82
animals
    transgenic, 225
anti-cancer drugs, 7
anti-human A, B or AB typing antiserum, 80
anti-viral drug ganciclovir, 235
antibiotic G418, 212 , 214
antibiotic resistance, 212
antibiotic-free media, 94
antibiotics, 3 , 23 , 94 , 178
antibodies, 129 , 134 , 138 , 178 , 211 , 233
antibody capture assays, 141
antibody secreting cells, 137
antibody-producing cells (B cells), 135
antibody-secreting cells, 150
antifoaming agents, 156 , 178
antigen, 138
antigen capture assay, 141
antigen specific hybridomas, 231
antigen-antibody reactions, 134
antigen-presenting cells, 139
antigenic shift, 136
antigens, 134
antimycoplasma monoclonal antibody-based kits,
148
Antonie van Leeuwenhoek, 2
applications in embryology, 198
aqueous humour, 49
arginase, 69
arginine, 69
artificial liver, 189
ascites tumours, 150

ascitic fluids, 49
ascorbic acid, 63 , 67
aseptic techniques, 3 , 23 , 148
asexual multiplication, 195
athymic mouse, 120
atmosphere, 61
ATP, 155
atropine-like drugs, 25
attachment and spreading factors, 67 , 68
autografting, 5 , 6
automated pipetters, 24
autoradiography, 78
avian erythroblastosis virus, 124
avian leukosis virus (ALV), 127
avian retroviruses, 228
avian tissue culture, 48
axons, 82
azaguanine, 144
azaserine, 145

B

B cell, 139
B cell tumours, 135
B cells, 37 , 59 , 136 , 138
        humans, 149
B type culture collection, 148
B vitamins, 61
bacteria, 22 , 70 , 94 , 140 , 207
bacterial cell wall, 229
bacterial contamination, 49 , 148
bacterial endotoxins, 69
bacterial origin of replication (*ori*), 212
bacterial plasmid vectors, 212
bacterial protoplasts, 229
baculovirus vectors, 232
baculoviruses, 231 , 232
balanced salt solution (BSS), 26 , 49 , 60
BALB/C strain, 137
basement membrane, 82
batch feeding of glucose, 155
batch technology, 178
benefit, 182
BHK21, 50
bicarbonates, 50
binding proteins, 69
biological activity, 178
biological fluids, 48
biomass, 164
biomass determination
        conduct methods, 36
        direct methods, 32
biopharmaceuticals, 167
bioreactors, 163

biostats, 178
birth, 191
Bishop, 122
bladder, 4 , 25
blastocyst, 9 , 191 , 193 , 200
blastomeres, 197 , 199
blood, 9 , 26 , 42 , 235
blood agar plates, 22
blood and lymph, 89
blood and lymph tissue, 83
blood cells, 25
blood clotting
        factor VIII, 208
blood group antigens, 80
blood products, 42
blood vessel, 235
bone, 4 , 25 , 82
bone marrow cells, 9 , 83 , 235 , 228
bone marrow transplant recipients, 6
Boveri, 121
bovine embryos, 199
bovine papilloma virus origin, 213
bovine serum albumin (BSA), 140 , 155
brain, 42 , 82
breast, 4 , 25
bunsen burner, 23
Burkitt's lymphoma, 127
burn wounds, 5
Burnett, 208

C

C-banding, 75
c-myc, 127
c-src, 122
caffeine, 195
Cahn, 36
calcium, 63 , 93 , 228
calcium phosphate transfection, 228
cancer, 6 , 116
cancer cell, 121
cancer diagnosis, 130
cancer metastasis, 116
cancer research, 4 , 129
cancer treatment, 235
capillaries, 37
carbohydrates, 140
carbon sources, 66
carboxymethyl cellulose, 65
carcinogens, 7 , 125 , 127 , 187
cardiac muscle, 4 , 82
Carrel, 3
cartilage, 4 , 25 , 26 , 82
categorising tissue culture, 189

cationic lipids, 229
cattle, 194
Caucasians, 77
CCAAT box, 216
cDNA, 209
cell banks, 25 , 108 , 110 , 148
cell body, 82
cell cloning, 15
cell culture, 3 , 8
cell death, 66
cell density, 178
cell division, 66
cell division and differentiation *in vivo*, 84
cell lines
    characterisation of, 74 , 75 , 77 , 79 , 81 ,
83 , 85 , 87 , 89 , 91 , 93 , 95 , 97
cell line, 12 , 29 , 80
cell line number, 29
cell lines, 50
    cloning, 36
    instability of, 101
    preservation of, 100 , 101 , 102 , 103 ,
    105 , 107 , 109 , 111 , 113
    propagation of, 29
    selection of, 36
cell lines from embryonic tissue, 190
cell polarity, 88
cell separation, 17
    based on affinity binding, 40
    based on density, 39
    based on size, 38
    based on surface charges, 39
    physical methods, 38
    using cytofluorometry, 40
    using flow cytometry, 42
cell shape, 88
cell strains, 12
cell surface antigens, 90 , 91
cell theory, 2
cell transformation, 116 , 118
cell type specific markers, 92
cell types
    characterisatioan of, 90
cell-cell contact, 86 , 87
cell-cell interactions, 93
cells, 9
    adherent, 61
    estimation of, 31
cells at low densities, 67
cellular material for growth, 164
cellular resistance to immune rejection, 122
cellulose sponges, 189
centrifuge tubes, 24
centromere region, 75

cerebral glioma, 235
characterisation of cell lines, 74 , 75 , 77 , 79 , 81 ,
83 , 85 , 87 , 89 , 91 , 93 , 95 , 97
characterisation of transfected cells, 232
checking media ingredients, 22
chemical instability of materials, 63
chemostats, 178
chemotherapy, 7
chick (and other embryo) extracts, 49
chick embryo extract, 187
chick plasma, 187
chicken cells, 13
chicken sarcomas, 122
chimaeric animals, 225
Chinese Hamster cells, 76
Chinese Hamster liver, 36
Chinese hamster ovary cells, 208
cholesterol, 69
choline, 67
chondrocytes, 68
chondronectin, 68
chromosomal analysis, 74
chromosomal damage, 66
chromosomal translocation, 127
chromosome analysis, 7
chromosomes in these photomicrographs, 76
cis-OH-proline, 16
Clark, 36
classification of oncogenes, 125
classification of proto-oncogenes, 125
cleavage of the 3′ end of RNA, 217
clone, 16 , 209
clone number, 30
clone-forming efficiency, 110
clone-forming efficiency test, 110
cloned genes, 206
cloning, 15
    by selective detachment, 37
    limiting dilution method, 146
    single cell picking, 146
    through interaction with a substratum, 37
cloning desired strains of domestic animals, 7
cloning efficiency, 14
cloning hybridomas, 146
cloning in soft agar, 147
cloning on feeder cells, 147
cloning rings, 37
clotted plasma substrate, 187
CNS tissue, 42
co-culture, 189
co-enzymes, 60
co-transfection, 231
CO_2, 24
CO_2-bicarbonate system, 155

coagula, 48
cobalt, 63 , 69
coconut milk, 49
cofactors in metabolism, 66
colcemid, 74
cold room, 24
collagen, 6 , 87 , 93 , 166 , 189
collagen coating on spheres, 167
collagenase, 28 , 37 , 167
colon carcinoma, 101
colony stimulating factors, 67
colony-stimulating factor-1 (CSF-1), 124
columnar, 82
commitment, 85
committed precursor cells, 85
complement components, 134
complement-mediated lysis, 16
complementary sequence, 221
complete medium, 155
complex media supplements, 182
concatemer formation and selection, 230
conditioned medium, 37
conductance, 35
congenital malformations, 200
connective tissue, 4 , 82 , 88
connexons, 87
consistency, 182
contact inhibition, 9 , 118
contamination
        intra-species, 78
contamination of the tissue cultures, 22
continuous cell lines, 4 , 231
continuous cultivation, 175 , 177
continuous flow bioreactors, 176 , 178
continuous flow cultivation, 173
continuous or immortalised cell lines, 13
controls of normal cell growth, 117
Coon, 36
copper, 69
copy (complementary) DNA (cDNA), 209
cornified cells, 88
Cortland, 60
COS cells, 213
Coulter Cell Sorter, 40
Coulter counters, 24 , 35
counting chambers, 24 , 32
covalent modifications, 207
Creutzfield Jacob disease, 42
cross contamination of cells, 22
cryopreservation, 103
cryoprotectants, 106 , 109 , 148
cryotubes, 106
cuboidal, 82

culture
        maintenance of, 29
        phenotypically heterogeneous, 92
culture collections, 111
culture flasks, 24
culture parameters, 154
culture vessels, 158
culture volume, 178
cultured fibroblasts, 88
cultured skin, 235
cultured vascular cells, 6
culturing cells, 24
cysteine, 66 , 68
cystine, 63 , 66 , 155
cytidine, 66
cytofluorograph, 40
cytogenetics, 74 , 78 , 127
cytokeratins, 91
cytokines, 182
cytokinesis, 84
cytomegalovirus (CMV), 6
cytoplasmic intermediate filament proteins, 91
cytoskeleton, 91 , 120
cytotoxicity testing, 7

D

de-differentiate, 86
DEAE sephadex, 166
DEAE-dextran, 208
defined media, 47
degeneration, 197
delivery of products *in vivo*, 235
denaturation of proteins, 65
dendrites, 82
determining sex, 201
dexamethesome, 68
Dextran T-500, 67
Dextran T70, 67
dhfr gene, 213 , 214
diagnosing inherited diseases, 78
diagnosis of genetic disease, 200
diagnostics, 130
dialysed serum, 49
dielectric constant, 35
differentiated cells, 85
differentiated state *in vitro*, 88
differentiated tumours, 117
differentiation, 2 , 67
        in vitro, 92
differentiation *in vivo*, 86
digestive system, 82
dihydrofolate, 213
dihydrofolate reductase (DHFR), 213

dilution cloning, 15
dilution rate, 177
dimethyl sulphoxide (DMSO), 106
diploid cells, 158
DIPSO, 51
direct cell interactions, 186
disposable pasteur pipettes, 24
dissection of embryos, 192
dissolution, 197
disulphide bonds, 207
division, 2
DMSO, 42 , 148
DNA, 36 , 209 , 228 , 231
DNA amplification, 127
DNA consensus sequences, 216
DNA fingerprinting, 78
DNA hybridisation, 148
DNA probes, 78
DNA stains, 22
DNA-dependent DNA polymerase, 209
dopamine, 6
double de-ionised water, 63
double stranded DNA provirus, 220
downstream processing, 176
downstream recovery, 182
draught tube, 173
Drosophila hsp 70 promoter, 216
drug efficiency testing, 129
drugs, 7
dulbecco, 50
Dulbecco's phosphate buffered solution A
        (PBSA), 61
dye exclusion test, 110 , 197
dyes, 42

E

Eagle's Basal Medium, 155
Eagle's medium, 57
Earle, 50
early transcription, 217
EC Commission, 93
EC-guidelines and Directives, 93
EGF, 67
eggs cells (ova), 199
electrofusion, 150
electronic particle counters, 35
        See also Coulter counter
electrophoresis, 39 , 233
electroporation, 229
elevation of vapour pressure, 65
embedded in a matrix of polyhedrin, 232
embryo splitting, 195
embryo culture, 191

embryo development, 197
embryo extracts, 3 , 48 , 60
embryo screening, 130
embryo transfer, 198 , 199
embryo transplantation, 7
embryo-derivated cell lines, 25
embryonic cell lines, 60
embryonic stem cells, 9 , 191 , 198 , 225
embryonic tissue, 9 , 190
encapsulating cells, 174
endocrine cells, 4
endocrine glands, 86
endocrine system, 82
endogenous genes
        modified, 225
endonuclease, 221
endothelial cells, 236
endotoxins, 63
energy, 164
energy for maintenance functions, 164
engineered retroviruses, 223 , 228
enhancers, 216
enriched medium, 37
enucleated oocytes, 196
envelope glycoprotein, 220
envelope proteins, 222
enzymatic activity, 233
epidermal cells, 67
epidermal growth factors (EGF), 67 , 124
epithelial cell clones, 189
epithelial cells, 25 , 37 , 68 , 90 , 101
epithelial differentiation, 86
epithelial growth factor, 88
epithelial lines, 80
epithelial precursors, 86
epithelial tissues, 4 , 82 , 87 , 88
epithelium, 27 , 82
epitope, 139
Epstein Barr virus (EBV), 149
erythroleukaemia cells, 89
erythropoietin, 208
essential amino acids, 63 , 66 , 155
essential fatty acids, 69 , 155
establised cell lines, 13
ethical controversy, 6
ethylenediamine tetra acetic acid (EDTA), 167
eukaryotes, 209
eukaryotic cells, 207
eukaryotic gene expression, 215
eukaryotic translation, 212
evaluation of embryo quality, 197
exchange of nutrients, 187
exonuclease, 221
explants, 11 , 187

exploding ampoules, 106
expression cassettes, 212 , 215 , 217
expression cloning, 211
expression of functional properties
      of the mature cell, 90
expression of the gene, 211
expression vectors, 211 , 215 , 231
expression vectors based on adenovirus, 213
expression vectors based on vaccinia virus, 213
external fertilisation, 193
extracellular matrix, 8 , 82 , 86 , 87 , 93 , 186 , 189

F

factor IX, 208
FAD, 66
fall armyworm, 232
fallopian tube, 199
fat, 26
fatty acids, 69
feeder cultures, 147
feeder layers, 37 , 88
Fell, 187
fermentation, 163
fermentor (bioreactor) systems, 163
fertilisation *in vitro*, 195 , 200
fertilised mouse eggs, 198
fertilised ova, 193
fertilised ovum, 84 , 191
fibroblast cell clones, 189
fibroblast cells, 101
fibroblast feeder cultures, 147
fibroblast growth factor (FGF), 67
fibroblast overgrowth in primary cultures, 70
fibroblastoid cell line, 101
fibroblasts, 4 , 27 , 37 , 67 , 82 , 90 , 147
fibronectin, 68 , 87
fibrous connective tissue, 25
Ficoll, 28 , 39 , 67
fingerprint, 80
finite cell lines or strains, 13
Fischer, 52
Fischer's medium, 59
fish, 60
floating raft, 188
flow cytofluorimetry, 17
flow cytometry, 17
fluid interface, 187
fluorescein isothiocyanate, 77
Fluorescence Activated Cell Sorter (FACS), 17 , 40
fluorescent dyes, 233
fluorochrome dye, 95
fluorochromes, 40
FMN, 66

foaming, 65 , 155
foci, 119
foci in *in vitro* culture, 118
focus formation, 118
focus forming assay, 119
foetal bovine serum, 37
foetal calf serum, 69 , 143
foetal stages, 191
foetal tissue, 190
foetuses, 191 , 197
folic acid, 59
follicle stimulants, 200
follicle stimulating hormone (FSH), 193
follicles, 193 , 199 , 200
foot and mouth disease, 4
forensic science, 78
formulation of media, 182
free radicals, 69
freezer, 24
freezing, 107
freezing point depression, 65
frog embryos, 89
functional assays, 141 , 143
fungal contaminations, 148
fungi, 22 , 94
fusing cells, 145

G

G-banding, 75
G11 banding, 76
G418, 224 , 231
gag, 222
gametes, 193
gammaglobulins, 69
ganglia, 82
ganglion, 89
gap junctions, 87 , 186
gas-lift (air-lift) type, 173
gaseous diffusion, 187
GC rich regions, 216
gel electrophoresis, 233
gelatine, 166
gelatine beads, 167
gene amplification, 127 , 213
gene expression, 209 , 211
gene manipulation, 206 , 209
gene probes, 130
gene replacement, 201
gene targeting, 7 , 224
gene therapy, 130 , 235
gene transfer vectors, 215
Gene-Probe, 96
genes resistance to antibiotics, 231

genetic disease, 121
genetic disorders, 235
genetic engineering
        medical applications, 234
genetic engineering of animal cells, 206 , 207 ,
209 , 211 , 213 , 215 , 217 , 219 , 221 , 223 , 225 ,
227 , 229 , 231 , 233 , 235 , 237
genetic instability, 101
genetic transduction, 122
genetically homogeneous, 92
genomic DNA, 209
genomic viral RNA, 220
gentic diseases, 198
gentically manipulating animal cells, 182
germline cells, 121 , 225
germline gene therapy, 235
Gey, 4
Giemsa dye, 75
glandular structures, 189
glass bead reactors, 163
glass distilled, 63
glass tubes, 160
glassware, 22
glial cells, 4 , 27 , 67 , 82 , 89
gloves, 42
glucocorticoid hormones, 68
glucose, 47 , 63 , 66 , 155 , 163 , 197
glucose 6-phosphate dehydrogenase (G6PDH), 77
glutamic acid, 155
glutamine, 63 , 66 , 155
glycerol, 102 , 106
glycoproteins, 82
glycosyl transferases, 207
glycosylate proteins, 207
glycosylation, 207 , 231
gonadotrophic hormones, 193
gonadotrophin, 193
gonadotropins, 193
Good Manufacturing Practice (GMP), 182
Gore-tex fibres, 189
government legislation, 206
Grace, 60
Grace's insect medium, 61
Gram negative bacteria, 63
Gram negative organisms, 94
Gram positive organism, 94
Gram stain, 22
Grid technique, 188
growth
        control, 121
growth factor concentrations, 167
growth factor receptors, 125
growth factor requirements, 118 , 120

growth factors, 37 , 47 , 67 , 68 , 69 , 125 , 129 , 155 ,
178 , 189
growth rate, 176 , 177
growth yield of a culture (Y), 165
growth-limiting nutrient, 176
guanosine, 66

H

haematopoietic system, 85
haemocyanin, 139
haemoglobin, 91
haemopoetic cells, 37 , 68
haemopoiesis, 89
haemopoietic (haematopoietic) lineages, 89
haemopoietic system, 235
handling tongues, 106
hanging drop cultures, 36 , 37
Hanks, 50
hapten, 139
Harrington, 208
Harris, 128
Harrison, 89
harvesting the cells, 167
HAT medium, 59
HAT selection, 144
Hayflick, 13
Hayflick effect, 13
hazards
        within the laboratory, 42
headspace (in a closed system), 155
Healy, 53
heat shock promoters, 216
heavy metal ions, 63
HeLa, 50
HeLa cell lines, 4 , 25 , 77
HeLa-cells, 178
helper virus, 224
HEPA, 23
heparin, 195
hepatitis, 208
hepatitis B vaccine, 208
Hepatitis viruses, 42
HEPES, 51 , 64 , 155
HEPPSO, 51
heritable retinoblastoma, 128
Herpes B viruses, 42
Herpes Simplex Virus, 223
Herpes Simplex Virus thymidine kinase, 235
heterogeneous cultures, 11
heterokaryon, 143
heterologous cell interactions, 189
heteronuclear RNA, 209
heteroploid cells, 158

high energy radiation, 125 , 127
high resolution 2-D electrophoresis, 80
historical development of organ culture, 187
histotypic culture, 189
HIV, 42 , 208
HL-A, 80
HLM, 50
Hoechst Dye (33258) staining assay, 148
holding medium, 30
hollow fibre cartridges, 174
hollow fibre culture, 161 , 175
Holmes and Scott, 60
Holtfreter, 60
homogeneous culture conditions, 169
homologous cell interactions, 189
homologous recombination, 224
hormonal control of the oestrous cycle, 193
hormonal environment, 189
hormonal factors, 67
hormones, 5 , 8 , 37 , 47 , 67 , 68 , 86 , 88 , 93 ,
129 , 155 , 186 , 187
host range of infectivity, 220
host system, 231
hot room, 24
HRPT, 144
human blood, 208
human cell culture, 13
human chromosomes, 76
human diploid cells, 157 , 167
human disease, 189
human embryo lungs, 101
human embryo viability, 200
human embryonic cells, 4
human embryos, 200
human epithelium, 80
human fibroblasts, 76
human foetal lung fibroblast cell lines, 9
human foetal lung fibroblasts, 13 , 190
human foetuses, 6
human growth hormone, 5
human hybridomas, 149
human infertility, 199
human interferon g, 5
Human Lymphocyte Antigens (HL-A) system, 80
human lymphocytes, 157
human menopausal gonadotrophin (HMG), 200
human monoclonal antibodies, 149
human platelet-derived growth factor (PDGF), 122
human reproductive technology, 199
human skin fibroblasts, 13
human/mouse hybrids, 76
hybridoma cell lines
        long term storage, 148
hybridoma colonies, 141

hybridoma lines, 167
hybridoma technology, 135 , 136 , 138
hybridomas, 5 , 59 , 70 , 76 , 134 , 135 , 137 , 139 ,
141 , 143 , 145 , 147 , 149 , 151 , 155
hydrocortisone, 37 , 68
hydrolytic nucleases, 228
hydroxylation, 207
hygromycin B, 231
hygromycin C, 212
hyperimmunisation, 140
hyperplastic cells, 116
hypotonic medium, 74
hypoxanthine, 59 , 144
hystatin, 94

I

identical twins, 199
identifying the stage of differentiation, 92
immortal cell lines, 10
immortalised B cells, 149
immune defence system, 120
immune rejection, 5 , 6
immune response, 137
immune system, 134
immunising mice, 138
immunoblotting, 233
immunofluorescence, 77 , 91 , 233
immunogen, 139
immunoglobulin genes, 137
immunoglobulins, 231
immunological assays, 97
immunological drift, 97 , 136
immunological markers on cells, 70
immunological tests, 77
immunosuppresion therapy, 6
impeller (propeller) type, 173
in vitro fertilisation, 7 , 193 , 197
in vitro genetic manipulation, 7
inactivation of enzymes, 65
increasing cell densities, 173
incubator, 24
induction of superovulation, 193
industrial cultivation of animal cells, 182
infertility, 7 , 200
inherited cancer susceptibilities, 128
inherited disease, 121
inoperable human brain tumours, 235
inorganic ions, 93
inorganic salts, 47
insect blood, 60
insect cell culture media, 232
insect cell lines, 60 , 231
insect cells, 24

insect haemolymph, 49
insects
        which feed on plants, 60
        feeding on animals, 60
insertional mutagenesis, 127 , 235
instability in cell lines, 101
insulin, 37 , 68
integration sites, 224
interband regions, 75
intercontinental shipping of embryos, 199
interferons, 5 , 161
interleukin 2, 37
intermediate filament proteins, 90
internal heating/cooling coils, 170
interspecies hybridomas, 149
intra-species contamination, 78
intracellular signal transducers, 125
intradermal (i/d) routes, 140
intravenous (i/v) routes, 140
introduction of the recombinant DNA, 227
introns, 209 , 217
invasiveness, 122
invertebrates, 193
inverted microscope, 24
inverted repeat, 220
isoelectric focusing, 80
isoenzymes, 76
isoenzymology, 76
isolation of eukaryotic genes, 211
isolation of the gene, 209
isoleucine, 66 , 155
isopycnic sedimentation, 39
isozymes, 76
IVF, 198 , 199
IVF in humans, 200
IVF technique, 199 , 200

J

jet-flow type, 173
Jolly, 3
Jones, 3

K

karyotype, 74
karyotype changes, 66
karyotype techniques, 74
keratinocytes, 88
keyhole limpet haemocyanin (KLH), 140
kidney, 4 , 25 , 187
killer T-cells, 6
Klein, 128
Knudson, 128

Kohler, 5 , 135
Kupffer cells, 36

L

labelled antibodies, 233
labelled DNA probes, 233
laboratory coats, 23 , 42
lactate dehydrogenase, 91
lactic, 155
lactose dehydrogenase (LDH), 77
lag phase, 12
lag phase of the growth cycle, 66
laminar air flow cabinets, 3
laminin, 68 , 87
laproscopy, 200
large scale animal cell culture, 154 , 155 , 157 , 159 ,
161 , 163 , 165 , 167 , 169 , 171 , 173 , 175 , 177 , 179 ,
181 , 183
large scale microculture, 167
large T antigen, 213
legal and ethical issues raised by IVF, 201
lens paper, 188
Leob, 187
lethally irradiated, 147
leucine, 155
leukaemic cells, 59
level specificity of DNA transcription, 215
lincomycin, 148
lipids, 63 , 67 , 69
lipofection, 229
liposome, 229
liquid nitrogen, 24 , 103 , 148
live, attenuated or killed cells, 140
liver, 4 , 6 , 25 , 182
localised proliferating tumours, 235
log phase, 164
logarithmic phase, 12
long terminal repeat (LTR)
        of mouse mammary tumour virus, 217
looped bioreactors, 172
low density cultures, 66
low plating efficiencies, 67
low serum, 70
LTRs, 224
lung, 4 , 25
lymph glands, 83
lymphatic system, 82
lymphoblastoid cells, 178
lymphocytes, 89
lymphoid cells, 127
lymphokines, 5 , 70
lymphomas, 70
lymphoprep, 28

lysophosphatidyl-choline, 195
lysosomal activity, 66
lysozyme, 229

M

macrophages, 37 , 89 , 147
maintenance functions, 174
maintenance medium, 30
major histocompatibility antigens, 80
major histocompatibility complex - MHC, 80
male pronucleus, 198
malic enzyme (ME-2), 91
mammalian cells, 24
mammalian vectors, 212
mammals, 193
mammary epithelial cells, 17 , 88
mammary gland, 189
manganese, 63 , 69
Mantegazza, 102
marker genes, 212 , 235
market authorisation, 182
Mary Warnock, 201
masks, 42
mated female mice, 192
materials used for large vessels, 169
Matthias Schleiden, 2
mature haemopoietic cells, 9
maturing oocyte, 199
maximum specific growth rate (μmax), 177
MCS, 224
        multiple cloning site, 214
measles, 4
measurement of key enzymes, 197
mechanical stirrers, 155
media
        and specific cell types, 70
        for cold blood vertebrate lines, 60
        for invertebrate tissues, 60
        low serum, 70
        nutritional components, 66
        selection of, 70
        serum-free, 70
media design, 46
medium, 155
medium 199, 52 , 60 , 66
medium 858, 53
medium perfusion, 156
medium reservoir, 175
medullary plate of a chick embryo, 187
melanocytes, 4
membrane phospholipids, 67
membrane reactors, 174 , 175
membranes, 176 , 197

metabolic intermediates, 60
metabolic requirements of embryos, 197
metaphase stage of mitosis, 74
metastatic spread, 116 , 122
methotrexate, 145 , 213
methylation, 207
methylcellulose, 67
methyl purine deoxyriboside, 96
Metrazamide, 39
micro-injected fusion genes, 198
microbial contaminations, 93
microbial infection, 178
microcarrier systems, 166
microcarriers, 167
microinjection, 229
microprojectiles (biolistics), 230
microscope, 22 , 24
microwell plates, 36
milk, 82
milk production, 17
milk proteins, 88 , 91 , 189
Milstein, 5 , 135
mineral oil, 137 , 143
minerals, 67
minimum essential medium, 57
Minimum Essential Medium (MEM), 61 , 155
Minimum Essential Medium Eagle, 57
mitogen, 117
mitomycin C, 147
mitosis, 84
mitotic inhibitors, 76
modulators, 122
molecular basis of cancer, 121
molecular genetics, 128
molecular probes, 22
Moloney murine leukaemia provirus, 223
Moloney murine leukaemia virus (M-MuLV), 228
molybdenum, 69
monoclonal antibodies, 5 , 16 , 70 , 77 , 91 , 135 , 208
monocytes, 89
monolayer dispersal, 61
Moorhead, 13
MOPC (mineral oil plasmacytomas), 137
MOPS, 51
Morgan, 52
morphology, 90 , 118 , 120
Morton, 52
Moscana's Ca^{2+} and Mg^{2+} free saline (CMF), 61
mouse 3T3, 125
mouse chromosomes, 76
mouse embryo fibroblasts 3T3 cell lines, 9
MRC-5, 9 , 25
MRC5 cell lines, 147
mRNA, 209

mRNA cap sites, 215
mRNA encoding viral proteins, 220
mucus, 82
mulitple market authorisation, 182
multi-stage carcinogenesis, 129
multicellularity, 2
multiple cloning site, 224
multitray units, 161
mumps, 4
murine leukaemia virus, 224
murine retroviruses, 228
muscle fibres, 89
muscle precursor cells, 89
muscles, 25
muscular tissue, 82 , 89
mutagens, 42 , 125
mutants, 14 , 101
mycoplasma, 22 , 49 , 70 , 77 , 95 , 148
mycoplasma DNA, 95
MycoTect, 96
myelin sheaths, 82
myeloma cells, 59 , 143 , 229 , 231
myelomas, 137
myoblasts, 27 , 89
myoblasts L6 cells, 89
myofibrils, 82
myosin, 82
myotubes, 89

N

N-linked glycosylation, 231
N$_2$ vapour, 106
NAD(P), 66
native conformations, 207
natural media, 48
NCTC 109, 53
Negroids, 77
neo, 214
neo gene, 212 , 214 , 224
neomycin phosphotransferase gene, 214
neomycin resistance, 235
neonatal mouse dorsal root ganglia, 89
neoplasms, 116
neoplastic cells, 116
nervous tissue, 82 , 89
nervous tissue of chick embryos, 191
Neubauer haemocytometer, 32
neural cells, 4 , 25
neural crest, 6
neural grafts, 6
neural transplantation, 6
neurofilaments, 91
neurons, 4 , 82 , 89

neurotransmitter, 6
new blood vessels, 236
Niacin, 66
non-nutritional medium supplements, 155
non-transformed cells, 118
normal cells, 118
normal human tissue, 13
normal life cycle of a retroviruses, 220
normal tissue, 10
Northern blotting, 233
nuclear inclusion bodies, 232
nuclear transcription factors, 125
nuclear transfer, 195
nuclear transplantation, 196
nucleated cells, 80
nuclei, 196
nucleic acid precursors, 66
nucleic acids, 140
nucleoside phosphorylase (NP), 77
nucleotidylation, 207
nude mouse, 120
nutritional components of media, 66

O

oestradiol, 68
oestrous cycle, 193
offspring, 197
oncogenes, 42 , 116 , 121 , 122 , 150
oncogenic, 228
oocytes, 193 , 200
organ, 82
organ culture, 8 , 25 , 186 , 191
organ replacement therapy, 182
organ systems, 82
organisation of cell in vivo, 82
organogenesis, 86
organoids, 189 , 236
organotypic culture, 189
origins of replication in animal cells, 213
origins of replication of viruses, 213
orthoclone OKT3, 97
osmolarity, 64
osmotic balance, 164
osmotic pressure, 47
ovarian stimulation, 200
ovary, 187 , 193 , 199
ovum, 199
oxidative phosphorylation, 155
oxygen, 156 , 175
oxygen supply, 171

P

packaging cell lines, 228
packaging lines, 224
packaging of viral RNA into virions, 223
packaging signal, 223
pancreas, 182
pancreatic islet cells, 4
paracrine factors, 86 , 93 , 186
Parker, 53
Parkes, 102
Parkinson's disease, 6
partial pressure of O_2, 156
partially complete synthetic media, 52
particulate proteins, 139
passage number, 12
passaging, 12
passaging cells, 12
Pateman, 36
patented cell lines, 111
paternity testing, 78
pBR322, 212
pBR322
      origin of replication, 213
PCR (polymerase chain reaction), 127 , 227
penicillin, 3 , 94
peptone, 155
Percoll, 39
perfusing medium, 175
perfusion cultures, 155
peripheral blood, 83
peripheral blood lymphocytes, 235
peripheral nerves, 82
peripheral nervous system, 89
peristaltic pump, 163
peritoneal cavities, 150 , 236
peritoneal cavity
     rat, 189
pH, 47 , 65 , 155
pH probe, 155
phagocytic cells, 134
phagocytosed, 228
phenobarbitone, 16
phenol red, 65
phenotype changes, 101
phosphate buffered saline (PBS), 30 , 50
phosphates, 50 , 155
phosphoglucomutases (PGM and 3), 91
phosphogluconate dehydrogenase (PED), 91
phospholipids, 69
phosphorylation, 207
pipettes, 24
pituitary, 4
plasma cells, 135

plasma clots, 48 , 187
plasmid vectors, 212
plasmid-based mammalian vectors, 215
plastic bags, 162
plastic film propagators, 162
plateau phase, 12
platelet-derived growth factor (PDGF), 67
pleural fluids, 49
pluripotent, 85
pluripotent stem cells, 235
pluronic F-68, 155
pneumonia, 6
pol II promoter, 215
polio vaccine, 4 , 180
poliomyelitis virus, 4
Polter, 137
polyacrylamide, 166
polyadenylation, 217
polyadenylation site, 217
polyamine oxidase, 69
polyaminoaldehydes, 69
polyclonal sera, 135
polyethylene glycerol (PEG), 143
polyethylene glycol, 229
polyhedrin, 232
polyhedrin promoter, 232
polymerisation chain reaction (PCR) technique, 233
polymers, 67
polymorphisms, 78
polyoma origin of replication, 213
polyoma virus T antigen, 213
polypeptide hormones, 89
polystyrene, 166
polyvinyl pyrollodine, 65
post fertilisation development, 195
potency, 182
pp60v-*src*, 122
pre-implantation diagnosis, 201
pregnant mouse, 192
preparation of cells for fusion, 143
preparation of primary cultures, 26
preparing cells for freezing, 106
preservation of animal cell lines, 100 , 101 , 103 , 105 , 107 , 109 , 111 , 113
primary cell culture, 11
primary cell culture medium, 60
primary culture, 29
primary cultures
     removal of non-viable cells, 28
primary monkey kidney cells, 4
primate species, 77
primer (tRNA), 222
prions, 42

production of embryos *in vitro*, 193
productivity of animal livestock, 198
progesterone, 68 , 193
promoters, 215
prostaglandins E, 69
prostaglandins F$_{2a}$, 69
protective clothing, 106
protein, 36 , 80 , 178
proteins
        of therapeutic value, 208
proteoglycans, 82
proteolytic cleavage, 207
proto-oncogenes, 121 , 122
protoplast fusion, 229
pseudopregnant females, 198
pSV2 plasmids, 217
pSV2*dhfr*, 214
pSV2*neo*, 212 , 214
pulse height threshold, 35
purine nucleotide synthesis, 144
purity, 182
purity of materials, 63
pyrrolidine carboxylic acid, 63
pyruvate, 197
pyruvate supplement, 66
pyruvic acid, 155
pZIPNEOSV(X)1, 224

Q

quality control assays, 182
quantitation of cell viability, 109
quinacrine dihydrochloride, 75
quinacrine mustard, 75

R

rabies, 4
radioactive isotopes, 233
radioactively labelled DNA probes, 78
radioisotope labelling, 36
random integration, 224
ras genes, 125
ras proto-oncogene, 125
rate of glucose utilisation, 164
Rb-1 cDNA, 129
Rb-1 gene, 128
recognition sequence for translation, 217
recombinant DNA, 206
recombinant DNA methods, 78
recombinant DNA molecule, 209
recombinant DNA technology, 121 , 182
recombinant proteins, 5 , 208

recombinant retroviruses, 224
        carrying a human gene, 235
recombinant retroviruses as genetic vectors, 150
reconstitution of damaged tissues and cells, 5
reconstruction of three-dimensional structure, 188
redox potential, 157
refrigerator, 24
regulatory issues, 182
regulatory sequences, 215
repair of damaged structures, 164
repeats, 220
replacement of damaged tissues and cells, 5
replenishment of stocks, 108
replicate cultures, 8
reproducibility, 187
research on human embryos, 201
respiration rate, 164
respiratory system, 82
restriction enzymes, 78 , 224 , 233
restriction fragment length polymorphism
        (RFLP), 78
retinoblastoma, 128
retinoids, 67
retroviral genome, 222
retroviral infection, 228
retroviral long terminal repeats (LTRs), 216
retroviral vectors, 218 , 228
retroviral-mediated gene transfer, 235
retroviruses, 122 , 125 , 218 , 223
reverse transcriptase, 209 , 220 , 222
riboflavin, 66
ribonuclease H, 220
Rinaldini, 61
ringer, 50 , 60
RNA, 209 , 221
RNA polymerase, 215
RNA polymerase II, 216
RNA processing, 209
RNA tumour viruses, 67
RNA-dependent DNA polymerase, 209
Robert Hooke, 2
Robinson, 187
rodent cell cultures, 14
rodent embryo cells, 14 , 190
rodent tissue culture, 13
roller bottles, 25 , 158
Rous, 3 , 122
Rous Sarcoma Virus (RSV), 122 , 228
route of inoculation, 140
Roux, 3 , 187
Roux bottles, 158
RPMI media, 59
rubella, 4
Rudolf Virchow, 2

S

safety, 182
    in the laboratory, 42
safety evaluation of cosmetics, 7
safety evaluation of drugs, 7
safety evaluation of food additives, 7
safety evaluation of industrial chemicals, 7
safety evaluation of pesticides, 7
sampling, 163
satellite DNA, 78
scale up, 178
Schneider, 61
Schwann cells, 82 , 89
screening hybridomas, 146
screening procedures, 140
secondary cell cultures
    from primary cultures, 29
secondary culture, 11 , 29
secondary oocyte, 199
seed stock, 100 , 101
selectable marker genes, 231
selectable markers, 212
selecting the appropriate bioreactor system, 178
selection, 14
selection for specific traits, 201
selective enrichment strategies, 227
selective inhibitors, 69
selective overgrowth, 101
selenium, 69
semi-synthetic media, 48
senescence, 101
sensitivity, 136
separating chimpanzee from human cell lines, 77
sera from newborn and adult animals, 69
serine, 66
serum, 4 , 27 , 47 , 48 , 60 , 65 , 67 , 69 , 135 ,
143 , 155
    components of, 68
serum substitute, 70 , 155
serum ultrafiltrates, 49
serum-free media, 67 , 70
settling out-decantation approach, 174
sex-specific antigens, 199
sex-specific DNA probes, 199
sex-specific nucleotide sequences, 199
sexing of embryos, 199
shear forces, 155 , 163 , 172 , 178
shuttle vectors, 213
silent gene regions, 226
silicone antifoaming agents, 65
silkworm cell lines, 60
Simian Sarcoma Virus (SSV), 122
Simms, 50

simple, 82
single cell cloning, 146
single point mutation, 125
single stranded DNA, 209
skeletal tissue, 4 , 82
skin, 4 , 25 , 182
skin biopsies, 27
skin grafts, 5
smooth muscle, 4
sodium carboxymethyl cellulose, 155
sodium dodecyl sulphate (SDS), 80
sodium ethylmercurithiosalicylate, 16
solid support, 187
solubility of materials, 63
soluble inducers, 93
somatic cell hybrid, 128
somatic cell therapy, 235
somatic cells, 121
somatomedins, 67
sources of cells for culture, 25
sources of embryos, 192
sources of tissue for culture, 24
Southern blotting, 78 , 127 , 232
soybean-casein digest, 94
Spallanzi, 102
sparging, 156 , 171 , 173
species verification, 74
species-specific antigens, 77
specificity, 136
spent medium, 155 , 174
sperm, 195
spermatozoa, 102 , 199
spermidine, 69
spermine, 69
spinal cord, 82 , 89
spinner flasks, 168
spinning method for fusing cells, 146
spiral film, 159
spleen, 9 , 143
spleen cells, 147
Spodoptera frugiperda, 232
sporadic retinoblastoma, 128
squamous, 82
ssDNA, 221
stability of the RNA transcript, 217
stable expression, 211
stable expression system, 213
stably transfected cell lines, 211
stacked plate reactors, 166
stage of differentiation, 81
static or batch cultures, 155
stationary phase, 12 , 164
steady state, 177
steam generator, 172

stem cells, 85
sterile dissection of the tissue, 11
steriliation of large vessels, 172
sterilisation *in situ*, 172
sterility
        checks, 48
steroid hormones, 217
steroids, 95
stirred cultures, 25
stirred reactors, 148
stirred tank bioreactors, 168
stirred tank reactors, 178
stirring method for fusing cells, 145
stirring rates, 171
stock solutions of salt mixtures, 52
storage in liquid nitrogen, 105
storage of embryos, 197
strains, 13
stratified, 82
streptomycin, 3 , 94
stroma, 82
stromal cells, 82 , 87
stromal precursors, 86
sub-culture of adherent cells, 3
sub-culturing, 12
sub-culturing hybridomas, 147
sub-genomic mRNA, 223
subcutaneous (s/c) routes, 140
substrate, 164
super-heated steam, 172
superovulation, 193 , 200
supplements batch tested, 69
surface aeration, 156
surface energy of air bubbles, 155
surface information exchange, 87 , 186
surface tension, 47 , 65
surrogate parenthood, 201
suspension cultures, 9 , 61 , 166 , 167
SV40 DNA, 217
SV40 early regions, 217
SV40 expression signals, 217
SV40 origin, 213
SV40 virus, 214
sweat, 82
synthetic media, 47 , 49
synthetic peptides, 140
Szybaski, 208

T

T cells, 37
tail bleeds, 140
targeted drug delivery systems, 130
TATA box, 215

temperature, 65
temperature control in large vessels, 170
temperature of storage, 105
tenascin, 87
teratological studies, 198
terminal repeated sequences, 222
terminal differentiation, 85
termination of transcription, 217
TES, 51
test bleeds, 140
test tube babies, 199
testosterone, 68
*tet*R, 212 , 214
tetracyclin resistance gene, 214
tetrahydrofolate, 213
thawing frozen cells, 109
Theodore Schwann, 2
therapeutics, 180
thermal shock, 107
thioglycollate medium, 94
three-dimensional culture techniques, 186 , 187 ,
189 , 191 , 193 , 195 , 197 , 199 , 201 , 203
thrombosis, 208
thymidine, 59 , 66 , 144
thymidine kinase gene, 223
thymus cells, 147
thymus gland, 120
thyroid, 187
thyroid hormones, 68
timing of birth, 201
tissue culture, 2 , 3 , 8
tissue culture grade reagents, 63
tissue culture media, 3
tissue disaggregation, 61
tissue extracts, 48
tissue grafting, 5
tissue matrix, 3
tissue plasminogen activator, 208
tissue specific gene transcription, 216
tissue specificity of DNA transcription, 215
tissues, 82
tissues that require enzymatic disaggregation, 27
toluidine blue, 110
totipotency, 84
toxic agents, 198
toxicity, 129
toxins, 7
trace elements, 69
traditional antibody preparation
        limitation of, 134
transcription, 209
transcription start site, 215
transcriptional start point, 217
transcriptionally active site, 224

transfected, 208
transfection, 125 , 209
transferrin, 69
transformation, 208
transformed cells, 9 , 68 , 101
transformed cells become
anchorage-independent, 120
transformation *in vitro*, 149
transgenes, 198
transgenic animals, 198
transient cloning expression system, 211
transient expression, 211
transient expression systems, 211
transiently transfected cells, 211
translation, 209
transport proteins, 67
tris (hydroxymethyl) aminomethane, 50
tRNA molecule, 220
trypan blue dye, 35 , 104
trypsin, 3 , 9 , 27 , 37 , 74 , 167
trypsinisation, 9 , 28 , 106
tryptone soya broth, 94
tumorigenicity, 14 , 120
tumour cells, 13 , 42 , 116 , 136 , 155
tumour infiltrating lymphocytes (TILs), 235
tumour suppressor genes, 122 , 127 , 128
tumour tissue, 10
tumours, 3 , 116, 118
two-hit theory, 128
tylosin, 148
type culture collection, 25
tyrode, 50
tyrosine, 63 , 66

U

undefined media, 47
urea, 80
uridine, 66
urokinase, 180
US Food and Drug Administration, 94
uses of embryonic tissue, 190
uterine cavity, 199
uterus, 192
UV facility, 24

V

v-*erbB*, 124
v-*sis*, 122
v-*src*, 122
vaccines, 5 , 129 , 180 , 208
    polio, 190
vaccines against influenza virus, 208

vaginal plug, 192
van der Eb, 228
variable chromosome complement, 14
variation in cell lines, 101
Varmus, 122
vascular endothelial cells, 189
vascular tissue, 6
vectors, 209 , 212
versene, 28
vesicles, 197
viability of cells, 35 , 104 , 174
vimentin, 91
vinblastine, 74
viral core, 220
viral core particles, 220
viral core proteins, 222
viral DNA, 208
viral infection, 228
viral life cycle, 208
viral particles
        bud, 218
viral plasmid vectors, 212
viral splice donor, 223
virally encoded env glycoprotein gene product, 220
virion genomic RNA, 220
virology, 4
virus particles, 232
virus titres, 167
viruses, 5 , 42 , 49 , 70 , 77 , 96 , 121 , 208 , 216 , 224
viscosity, 47 , 65
vitamin A, 93
vitamins, 47 , 60 , 66 , 93 , 155 , 187

W

watchglass technique, 187
water, 63
water jacket, 170
waterbath, 24
Western blotting, 233
whole serum, 49
WOP cells, 213
work area, 23
working stock, 100 , 101
Wyatt's medium, 60

X

X-rays, 125

Y

Y chromosomes in humans, 75
yeasts, 94 , 140 , 207

Z

zinc, 63 , 69
zygotes, 191 , 197 , 199